A Multidimensional Life

The Legacy *of* Arjun K. Gupta

A Multidimensional Life

The Legacy *of* Arjun K. Gupta

As told to Laura Duggan

A Multidimensional Life:
The Legacy of Arjun K. Gupta

©2022 Arjun K. Gupta
All Rights Reserved

All rights reserved. No part of this publication may be reproduced, stored in a retrieval system, or transmitted in any form or by any means (electronic, mechanical, photocopying, recording, or otherwise) without the prior written permission of the author and the publisher.

Published by: Nicasio Press,
Sebastopol, California
www.nicasiopress.com
Cover Design and Photo Gallery by Constance King Design

ISBN: 978-1-7375814-3-7 (hardback edition)

Dedicated to my parents, Amar Nath and Leela Gupta, with love
—Arjun Gupta

Table of Contents

Author's Note 3
Glossary of Names 5
Timeline 9

PROLOGUE 1

PART I: INDIA – THE EARLY YEARS 5
    1. The Gupta Lineage 7
    2. The Companionship of Siblings 15
    3. Knowledge is Power 29

PART II: AMERICA – A NEW VISTA OPENS 51
    4. In Pursuit of a PhD 53
    5. Post-Graduation: New Family, New Locations 75
    6. Professor Gupta 99

PART III: THE WORLD - TRAILBLAZER FOR STATISTICS 137
    7. A Scholar with Heart 139
    8. An International Ambassador of Statistics 159
    9. Reflections 205

APPENDIX A: EDITORIAL ACTIVITIES 217

APPENDIX B: DR. GUPTA'S PhD STUDENTS 219

APPENDIX C: RESEARCH COAUTHORS 223

APPENDIX D: PUBLISHED BOOKS 227

APPENDIX E: INTERNATIONAL ASSIGNMENTS 229

APPENDIX F: END NOTES 231

Author's Note

The privilege of writing about someone's life, telling their story accurately and honestly, is both an honor and a challenge. It is an honor to be brought into someone's interior world of recollections, memories, feelings, and details. At the same time, it is a challenge to fully capture in words a multifaceted person such as Dr. Arjun K. Gupta. Therefore, much of this book is in the form of his own words and the words of those who know him best—his family, friends, and colleagues. Through the generous time that each person, most especially Dr. Gupta and his family, gave to the project, this narrative has taken shape.

Over a period of about six months, I had the privilege to conduct oral interviews with Dr. Gupta as well as with his colleagues, friends, and family. During this time, it became apparent that his accomplishments in the world of statistics and the scope of his international travels represented a story that needed to be shared. He acted as a mentor to many aspiring students, and his own efforts provided a model for many others to emulate. Yet despite his accomplishments, he never seemed to regard his efforts as unusual. This matter-of-fact demeanor is in strong contrast to the high esteem in which others hold him. The more I learned of his life, the more inspired I became to capture as much as possible. Dr. Gupta's journey from a small town in India to the world-stage of statistics demonstrates how much is possible with commitment and determination. His dedication to education and family are a light for future generations, for whom this book is written.

Acknowledgments

This book would not be possible without the unwavering commitment of time and interest from Dr. Gupta's family. Alka Gupta, Mita Gupta, and Nisha Gupta along with their mother,

Mrs. Meera Gupta, were the driving force behind this project. They offered hours of their time to review the text, give suggestions, arrange meetings, and set up interviews. Others who offered their valuable time for interviews included Dr. S. K. Bansal, Dr. W. J. Conradie, Ms. Jyoti Gupta, Dr. Pradeep Gupta, Dr. Aridaman Jain, Mr. Pankaj Manglik, Air Commodore Vishwanath Prakash, Dr. Jen Tang, and Dr. Harender Vasudeva.

Most especially, my gratitude goes to Dr. Gupta, for his good-natured, patient, and astute participation in the project, sharing his recollections and offering feedback. I have been personally uplifted and enriched through Dr. Gupta's presence. I hope that through this book, readers have their own experience of this remarkable and loving human being.

Laura Duggan

Glossary of Names

This story of Dr. Gupta's life is told mainly through the words of others. For easy reference, those who are mentioned or quoted frequently are listed below. Terms in italics signify the terms that Dr. Gupta uses to refer to his elders, and out of respect, this book has adopted that usage.

Arjun Gupta's Family

Mrs. Meera Gupta
    Dr. Gupta's wife.
Shrimati Manbhi Devi
    Grandmother; *Ma*.
Shri Lalita Prasad Gupta
    Grandfather; *Baba*.
Shri Amar Nath Gupta
    Father; *Pitaji*.
Shrimati Leela Gupta
    Mother; *Bhabhi*.
Shrimati Shakuntala Devi
    Eldest sister; *Bibi*.
Dr. Ram Nath Gupta
    Eldest brother; Bhaiya.
Mrs. Jyoti Gupta
    Niece; daughter of Ram Nath.
Mrs. Sarla Prakash
    Elder sister; *Jiji*.
Air Commodore Vishwanath Prakash
    Nephew; son of Sarla.
Mr. Sushil Manglik
    Elder brother; *Sushil Bhaiya*.
Mr. Pankaj Manglik
    Nephew; son of Sushil.

Mr. Vinod Kumar Gupta
: Youngest brother.

Ms. Alka Gupta
: Eldest daughter; married to Sharad Rastogi.

Arhaan Gupta-Rastogi
: Grandson; son of Alka and Sharad.

Ms. Mita Gupta
: Second daughter.

Ms. Nisha Gupta
: Third daughter, married to Patrick Nadol.

Saamik and Anika Nadol
: Grandson and granddaughter; children of Nisha and Patrick.

Family Friends, Classmates, Colleagues, and Students

Dr. Jim Albert
: Colleague at Bowling Green State University (BGSU).

Dr. Sabri Al-Ani
: Purdue classmate and host in Iraq.

Dr. Alphonse Amey
: Former student and host in South Africa.

Dr. S. K. Bansal
: Family friend and physician.

Dr. Hamparsam Bozdogan
: Former student and colleague; host in Turkey.

Dr. John Carson
: Former student.

Dr. John Chen
: Colleague at BGSU.

Dr. W. J. Conradie
: Colleague; host at University of Stellenbosch, South Africa.

Dr. William A. (Bill) Ericson
: Former department chair at University of Michigan.

Dr. Subash Goel
: Family friend; University of Michigan.
Dr. Zakkula Govindarajalu
: Family friend and colleague; University of Kentucky.
Dr. Pradeep Gupta
: Friend and Purdue classmate.
Dr. Aridaman Jain
: Friend and Purdue classmate.
D. K. Nagar
: Former student; host in Colombia.
Dr. Jen Tang
: Former student and research collaborator.
Dr. Tamás Varga
: Former student.
Dr. Harender Vasudeva
: Friend and professor of English at BGSU.

Notable Statisticians whom Dr. Gupta met or knew personally

Dr. Shreeram S. Abhyankar
: Faculty member at Purdue and family friend.
Sir Ronald A. Fisher
: Father of modern statistics.
Dr. V. S. Huzurbazar
: Former department head, University of Poona.
Dr. Eugene Lukacs
: Former faculty BGSU.
Dr. K. C. S. Pillai
: Arjun Gupta's advisor at Purdue.
Dr. C. R. Rao
: Luminary in the world of statistics.
Dr. Shankar Shrikhande
: Former professor at Banaras Hindu University.

Timeline

1938	Arjun K. Gupta born
1950-54	High School at Muzzafarnagar
1954-55	Intermediate College, Banaras Hindu University (BHU)
1955-59	College, BHU
1959	Graduation, BSc, honors, BHU
1960-62	Graduate student, University of Poona
1962	Professor, Agra University
1963-67	PhD student, Purdue University
1967	Marriage to Meera Nath, Dec. 25, Delhi
1968	PhD awarded, Purdue, January 1968
1968	Assistant Professor, University of Arizona
1970	Birth of first daughter, Alka, in March
1971	Assistant Professor, University of Michigan
1971	Birth of second daughter, Mita, in October
1976	Birth of third daughter, Nisha, in April
1976	Associate Professor, Bowling Green State University (BGSU)
1978	Full Professor, BGSU
1985-87	Department Chair, BGSU
1990	First Eugene Lukacs Symposium at BGSU
1999	Honored as Distinguished University Professor
2001	United States Citizenship
2015	Dr. Gupta retires from BGSU faculty

Prologue

Outside, it was almost 80⁰F, unseasonably warm on November 5, 1999 in Durban, South Africa. But inside the completely full lecture hall, it was comfortably cool. All eyes were directed toward the podium as the keynote speaker was introduced. A tall, sturdy-looking, handsome man with thick, white hair fringing his shining bald head stepped forward with unfeigned dignity and confidence. Dr. Arjun Gupta adjusted his glasses, offered a slight smile, and began to deliver his address to an international gathering at the South African Statistical Association's annual conference. Despite the technical density of the subject, for thirty minutes no one stirred, rapt as they followed the contours of the talk. When the talk concluded, a small commotion erupted as people politely jostled for a place in the line that formed to meet Dr. Gupta in person, a line that ultimately snaked out of the conference room itself. Some were carrying a gold-covered book, *Elliptically Contoured Models in Statistics,* the third edition of his book as well as the title of that day's lecture. Others carried his classic book, *Advances in Multivariate Statistical Analysis: Pillai Memorial Volume,* which he had created to honor his mentor and advisor, Dr. Pillai. Yet others lined up just to meet this legendary statistician in person.

Sitting in the back of the hall that day were Dr. Gupta's wife, Meera Gupta, and one of his daughters, Mita. It was a pivotal moment for Mita, who later shared, "As he was finishing, people lined up to get his autograph on his book, and to meet him and speak with him. It was like he was a rock star! We had never seen anything like that before." Seeing the admiration of this very distinguished group of academics toward her father, all the years of observing his dedicated work to teaching and statistics suddenly made sense to her.

One reason this enthusiastic reception came as a surprise to his family was Arjun Gupta's self-effacing humility. Even his closest friends admit they never really knew the importance of his statistical work until someone else pointed it out. Dr. Harender Vasudeva, a close friend and colleague at Bowling Green State University, shared at Dr. Gupta's retirement party, "I've known Arjun since '76 when he came here, and we have been friends since then. He has been just like any one of us, but I didn't know I had a friend who was a genius. I have not been able, even till today, to guess how much he has achieved in the field of statistics. It is going to be hard for me to treat him as a normal friend, because he is exceptional."

Exceptional is certainly an apt word for Dr. Gupta's contribution to the world of statistics. He has published over 530 papers, which averages almost ten each year over the span of his fifty-three years of teaching. He has advised more than thirty PhD students, who themselves have over forty-six students spanning six of the seven continents of the world. He has traveled to more than forty countries presenting his research and teaching statistics. His research in the statistical subject known as Wilks' Lambda is still in use today, forming one of the basic tables used for multivariate statistical test analysis and machine learning.

This exceptional outpouring of work comes directly from Dr. Gupta's dedication to his craft. He exemplifies a man who paid his dues, so to speak, by overcoming all the challenges life presented him with a somewhat stoic attitude. His philosophy is best captured by his comments on how to handle adversity: "Life is nothing but hard work. Don't give up. Don't give up."

Working hard is not just reserved for his profession in statistics. Equally important in Dr. Gupta's world-view is working diligently on one's education. When asked to give advice to future generations, he replied, "Education, Education, Education." He himself has been a lifelong learner, keeping up with his field of statistics by reading journals even on the plane as he traveled to be a guest lecturer at universities around the globe. Education always comes first in his mind. His nephew Pankaj shared about his uncle, "*Chachaji* always said two things about education. One, it never goes to waste. It doesn't matter what you're studying, as long as you are studying. He felt very strongly that education was foundational. And second, just doing a bachelor's was not enough. That was just his mindset. And he's passed it on to me, and I'm sure to other people."

To find the source of these twin themes of dedicated work and education, it is helpful to travel back to a small town in the state of Uttar Pradesh, India, and meet Arjun Gupta's parents, grandparents, and siblings.

Part I: India – The Early Years

No one can take your education away from you.
                                              Arjun Gupta

A MULTIDIMENSIONAL LIFE

~ CHAPTER ONE ~

THE GUPTA LINEAGE

Closely examining the front door of a three-story house outside the village of Purkazi, one can still see the carved name *Lalita Prasad Gupta*. The carving dates back to around 1890, when Arjun's grandfather was gifted the house, the land around it, and three villages by the British government. Offering land grants as an acknowledgment of service was a common practice by the British government at that time as a way to secure its imperialistic footing in India. By appointing local Indians as *zamindars*, or land holders, the British could collect tax without directly overseeing and managing the land. Although the zamindar system was abolished after India won its independence, it served as the backdrop for young Arjun Gupta's early life.

Upon receiving the gift of land and appointment as zamindar, Arjun's grandfather Lalita Prasad, who had served the British as a deputy transport manager, now found himself as a landlord and manager of a vast rural estate in the agricultural district known as Muzaffarnagar, in the state of Uttar Pradesh in northern India.

The Gupta family home initially housed Lalita Prasad Gupta, his wife Manbhi Devi, and their two sons and a daughter. Their daughter eventually married and moved in with her husband. The younger son obtained a government job as a teacher in another city. That left the eldest son, Amar Nath Gupta at home. Arjun's grandfather was a big supporter of education and sent Amar Nath to college in Dehradun for a few years, even though his son would eventually take over the role of zamindar.

When Amar Nath returned, he married Leela Gupta who, following the Indian custom of that time, moved into the home of her husband and his parents. It was in this extended family home in Purkazi that Arjun Kumar Gupta was born on March 25, 1938, the fifth of the six children of Amar Nath and Leela Gupta.

Initially, the lands given to Arjun's grandfather were only gardens, lacking any trees, but over time, his grandfather turned the land into a major agricultural enterprise. The climate in Purkazi was quite favorable for agriculture. It was hot in the summer and cold in the winter—though it was nothing compared to the cold weather that Arjun later encountered in Michigan, Indiana, and Ohio. There was a monsoon season, but it was not severe enough to flood the roads, so Arjun could still walk to school in the rainy season. However, the climate provided the perfect environment for growing crops. Arjun's grandfather arranged to plant hundreds and hundreds of trees in that very fertile soil, creating thriving mango and lychee farms. Eventually, truckloads of mangos and lychees were dispatched from the Gupta fields to the district and more distant commissaries. The newly-developed farms became a thriving business endeavor, offering a comfortable income for the Gupta family. Dr. Gupta described life with his grandfather.

> My grandfather, who I called *Baba*, used to tell us very good stories, often Urdu stories. After dinner we would go to bed, and he would tell us a story, or read a story from a book called *Alif Laila*, an Urdu book based on the *Thousand and One Nights*, or the *Arabian Nights*.

Arjun's grandmother, Manbhi Devi, or *Ma*, was the matriarch of the house, loved and respected by everyone. Ma was a very loving and affectionate person, always giving young Arjun hugs, especially when he came home from school. "I still miss her hugs. Even today, I cannot forget them," he shared decades later. An old photo of his grandmother shows her sitting in the interior courtyard of the Purkazi house, holding her *mala*, or prayer beads, that she used to recite her prayers. Her devotion influences Dr. Gupta even today, as he too uses his mala to recite prayers before going to bed. Meera Gupta described his morning ritual: "He has a little temple to put the deities in, and every morning, he will light incense and a *diya* (light)," a practice he learned from his beloved grandmother.

Recalling his grandparents, Dr. Gupta shared, "My grandfather and my grandmother are very vivid in my mind. Growing up with them was the best time for me."

While affection and devotion were qualities that Arjun received from his grandmother, he received complementary traits from his mother: generosity and open-mindedness. Leela Gupta, respectfully called Bhabhi by the children, was a very generous woman—no matter who came to the door, she would offer them something. Dr. Gupta's niece Jyoti Gupta tells the following story:

> One time, his [Arjun's] mother was cooking when an old *sadhu*, a holy man, came by and asked for food. She fed him, showed him a lot of respect, and gave him money and clothes. He granted her a boon that if she got bitten by a snake, the venom was never going to affect her. She did get bitten several times, and she never got sick from it.

People of all religions were welcomed into the Gupta home by Bhabhi. Whether they were Muslims, Hindus, or Christians, everyone was treated with the utmost respect. As Dr. Gupta recalled, "There were Christian missionaries up there too. There was one who was always a welcome guest in our house. He was just a very nice person." This welcoming attitude served Dr. Gupta well in later life when he traveled the world, interacting with people of every culture and religion including Muslims in Arab countries, Buddhists in Asia, and Christians in Africa and South America. He shared, "My ability to travel to so many different countries was surely because of my mother. She was very open-minded."

All the Gupta siblings loved their mother, who inspired each of them to grow beyond the confines of the town, which was an idea somewhat radical for that time. But Arjun, perhaps because he was the third son to leave home, was especially close to his mother, and her generous spirit is strongly reflected in his character. Dr. Gupta's home in Bowling Green has always been open to everyone—colleagues, students, relatives—whether for meals, visits, overnight, or even extended stays. His friend Dr. Vasudeva observed,

> When Arjun meets with people, he is very warm and friendly, welcoming. He has played the role of being a kind of senior statesman for the Indian origin community. He wants to know everybody who has come and invites them to his house. He and his wife Meera are hospitable in that regard.

As was the custom, Arjun's father inherited the role of zamindar, managing and administering all the land and houses in the three villages. He carried himself with great dignity, as befitted a significant landowner in town. Sometimes, Arjun accompanied his father, whom he always called *Pitaji*, as he walked the half-mile distance from the family compound to the mango fields to oversee the activities. Along the way, the villagers always greeted them with great respect. The villagers looked to Pitaji to support them, financially and otherwise, as that was expected of the zamindar. Yet

Pitaji was self-effacing and didn't consider his role as something big but rather simply his duty.

Pitaji looked like a landowner in the way he walked and the way he dressed. When he went to the mango fields to check on things, he could be seen wearing a cap and his double-breasted jacket. Dressing well was definitely one of the traits Arjun and perhaps all the sons inherited from their father. Later in his life, Dr. Gupta was rarely seen without a tie, even when relaxing on a holiday. His nephew Pankaj mused about this.

> Even when they visited the village, my dad always wore a tie and a coat, and Chachaji [uncle] was always very well dressed. That was the one thing that I never understood, because they came from a really small village yet they were very Westernized in the way they dressed.

Dr. Gupta describes his father as the epitome of a kind, self-sacrificing father who did everything he could for his children and other people. At the same time, Pitaji was somewhat traditional.

> My father was a strict disciplinarian. In Indian families, the father's word is the final word to a great extent, and that was certainly true for my father.

Perhaps that is one of the traits that didn't get passed on, because he admitted, "With my daughters, we changed that pattern as they grew up."

Dr. Gupta's devotion to education comes directly from his father. He shared, "Pitaji was always for education; it was his first preference for us." He added that at one point, his father told him, "There are only two things you will need: education and muscle power. If communism came to India, you would have no *seva*, no work to do. So education is the only thing."

Each of the adults in this extended family—father, mother, grandfather, and grandmother—left unique and indelible marks on Arjun's character. He continues to credit them with all that he achieved and his ability to overcome obstacles.

> The inspiration that you get from your parents, or your brothers and sisters and others, is very important. In my case, I think my parents were the guiding force. I could invoke them in my memory and thoughts of them always helped me through things. That influence was very powerful, and it lasted me eighty-two years. It is even with me today.

In 1954, just before Arjun turned sixteen, his grandfather passed away, followed about two years later by the passing of his warm-hearted and loving grandmother. The absence in the family home was palpable—no more stories after dinner from Baba, no more hugs from Ma.

Around this time, Arjun went away to high school in the city of Muzaffarnagar, joining his brother Sushil Bhaiya who was already there. When the two brothers came home every two weeks, it was to a dramatically changed family configuration with only his parents and youngest sibling Vinod at home in Purkazi.

A few years later, while Arjun was away at school, his mother became ill with undiagnosed ovarian cancer, and his father began to feel the effects of high blood pressure. Only four years after losing his grandparents, both his mother and his father passed away. Losing everyone within a few years of each other was a sad, almost devastating time for everybody, shaking the family structure. Since Ram Nath (*Bhaiya*), the eldest son, was overseas finishing his medical training when his mother passed away, Arjun returned home to take care of the rites for his mother.

By the time his father passed, Arjun was teaching in Agra, so he came back to Purkazi to handle things. "It certainly was a sad time. Because my younger brother and I had to take care of everything, more or less—the land, the houses—it was really difficult," Dr. Gupta shared.

~ Chapter Two ~

The Companionship of Siblings

The supporting structure of Arjun Gupta's life could have been severely shaken by the loss of the older generation had it not been for his unshakable connection to his siblings. Arjun's dedication to his family now found its fullest expression in his relationships with his five siblings and an extended network of spouses, nieces, nephews, and ultimately children and grandchildren. He shared, "It was wonderful to have the support of my siblings growing up and even later in life. Otherwise, I'd be lost. I certainly did and do miss my family—my brothers and sisters. We were very close."

Each of his siblings—two older brothers, two older sisters, and one younger brother—held unique roles in Arjun's life. Yet the common feature among all the relationships is dedication and a way of caring for each other that transcends time and place. Dr. Gupta's lifelong love and dedication to family is readily apparent when he recounts stories of each sibling.

The Eldest Sister

With the passing of his parents and grandparents, his eldest siblings took on even greater significance in Arjun Gupta's life. "My eldest brother and sister were very involved in my education and marriage. They were like parents because by the time I made these decisions, my parents weren't there anymore."

The eldest sister, Shakuntala Devi, or *Bibi*, as she was respectfully called, was twenty years older than Arjun. She was married two years before Arjun was born and as custom dictated, she left Purkazi to live with her husband, an accountant in the Royal Air Force. Despite moving away, she maintained close connections to the Gupta family and to her baby brother Arjun.

In the early 1940s, Bibi's husband was stationed in Karachi, and since it was part of India at that time, his whole family was stationed there as well. The extensive British railway system made it possible for Bibi to travel from Karachi to visit the Gupta family in Purkazi from time to time, where she saw Arjun grow from a baby to toddler to a lively young boy. By mid-1940s, however, life began to change in Karachi.

During India's independence movement, which is well-documented in books, movies, and stories, a tragic division between Muslims and Hindus festered and grew, despite India's long history of religious tolerance and nonviolence. As one author put it, the British motto was "Divide and Rule," which they found easy to do between the two religious groups.

Dr. Gupta shared, "Originally in Karachi, everyone got along: Hindus and Muslims. But after the 1947 partition, it was difficult even to cross the street. You were never sure whether you would come back alive or not." By December 1947, three months after independence, no less than four million people had been evacuated from what overnight had become Pakistan. One of those people was his sister Bibi who, along with her husband and four children, fled Karachi, leaving everything behind. Her four children ranged in age from two to nine years.

Train travel during that time was hair-raising, and one never knew if the train would make it through the turmoil to reach India without being blown up by fighting on either side. Yet, as Dr.

Gupta later shared, Bibi and her family had to risk the train, because Karachi was no longer safe for them. "Just before leaving, my sister sent a photograph of her children, herself, and her husband, saying, 'If we don't reach you, this is it.' It was heartbreaking." The family boarded a train to travel to Purkazi with few belongings other than the clothes they were wearing.

The train trip was somewhat circuitous involving several transfers but eventually, the family reached the nearby city of Muzaffarnagar. Bibi and her travel-weary family gratefully disembarked the train into the waiting arms of her father, who received them in an emotional reunion. They climbed into a local bus for the last thirty kilometers of their arduous journey to safety. As the familiar family house in Purkazi with its lush mango orchards came into view, Bibi teared up with relief. Nine-year-old Arjun hugged his sister as she arrived in the house, barely registering the fact that she and her children were now refugees in his parents' house. Arjun quickly befriended Bibi's eldest daughter Pushpa, who was almost exactly his age, embracing her as a playmate and sister, although Pushpa is technically his niece.

Eventually the newly formed government of India took responsibility for the refugees, establishing a program to find them jobs, and resettling those who had fled Pakistan. Bibi's husband was offered a job in Delhi, and the family moved there to painstakingly rebuild their life.

After Arjun's mother passed away, Bibi, as the eldest daughter, took on the role as the dominant woman in the extended Gupta family. Dr. Gupta's niece Jyoti explained,

> I remember when my parents were talking about getting all these boys married, my dad said that my dad's older sister would be the one who would preside and really say, "This is the custom. This is what happens. This is how we do it in

our family." So she [Bibi] had that kind of very, very dominant role in the family.

All the Gupta siblings were very close to their eldest sister, and Arjun demonstrated this by faithfully visiting her during each trip to India. Dr. Gupta's model of sibling love has created a very strong web of caring and support among the multi-generational, geographically dispersed family of aunts, uncles, cousins, and siblings.

The Eldest Brother

It was the year 1928. From the birthing room in the Purkazi house, a cry was heard, and shouts of *"Jai, Jai"* erupted as Leela Gupta at last delivered her first baby boy, Ram Nath. The house echoed with *"Betaa hu-a hai, betaa hu-a hai! Ladoo baato, ladoo baato!"* "It's a boy, it's a boy; distribute sweets, distribute sweets!" The first-born son holds a very important place in the structure of traditional Indian families, and Arjun Gupta's eldest brother, whom he always respectfully called *Bhaiya*, held that role in the Gupta family constellation. Bhaiya's daughter Jyoti shared a vivid illustration related to her father, based on a custom in India at the time. "Every year, he would be weighed on the scales against rice and grain. Whatever the weight was, that amount of grains and rice would be distributed among the poor."

Bhaiya was the first of the four Gupta boys to benefit from their father's commitment to education. Pitaji encouraged his eldest son to become a doctor and sent him to Lahore for intermediate college education. Bhaiya then went to King George's Medical School in Lucknow. Finally, he was educated abroad, first for a four-year residency at Vanderbilt University in Nashville, Tennessee, then in England for the Member of the Royal College of Physicians (MCRP), the postgraduate diploma in cardiology. In total, he was away from India for six years.

His elder brother's departure overseas for education made a deep impression on young Arjun, who was only about twelve years old at the time. Observing firsthand his father's sacrifices to educate all of his sons, Arjun Gupta also prioritized education. His choice of a career as an educator certainly has roots in his father's actions.

Sending my eldest brother overseas for school was a big deal at that time, partly because it was such a big expense. My father made a lot of sacrifices to send all the boys to school, to college, and beyond.

His father's sacrifice was both financial and emotional. Including transportation expenses, it could cost thousands of dollars to send even one child overseas, enough for the full education of at least two or three people in India. In addition to the financial concern, sending a son out of India at that time was also a major cultural concern. One of Dr. Gupta's college classmates, Dr. Pradeep Gupta, described it like this:

On one hand, everybody in India met somebody who came back from the US after education. They respected him and wanted to talk to him, so there was attraction. At the same time, they were afraid that their own family member going to America would change in some negative way. It was attraction and repulsion at the same time. If you were the first one in your family, it would have been quite a struggle. My father was thinking if I went to America, met an American girl, and got married, what would happen to the family? He had those kinds of qualms. My mother, on the other hand, said, "No. I believe him. I trust him. He will come back."

It is a testimony to Arjun's parents that they were both supportive of their eldest son going overseas to study, and this open-minded attitude was a great benefit when it later came time for Arjun to make a similar decision for himself. Nevertheless, it was a huge decision for the parents. Dr. Gupta shared, "When Bhaiya was leaving, my father came to see him off. It was kind of heartbreaking for my father."

Eventually Bhaiya became the chief medical officer in the Indian Railways, transferring to hospitals all over India, from Calcutta to Nagpur and Bombay, finally settling in New Delhi around 1978. His spacious and accommodating house became the site of the many family gatherings that pulled Arjun Gupta back to India every two years.

Bhaiya became like a parent to Arjun after their parents died, encouraging his younger brother to go overseas to study and later, arranging his marriage. Arjun always showed the utmost care and respect to his elder brother, who was ten years his senior. Bhaiya's daughter Jyoti described a typical interaction between her father and her uncle when the two of them met.

> When my dad would sit down, Chachaji would pull out the chair for him, and make sure he was comfortable. He would open the door for him. He didn't have to, but he did. This is very Indian. That respect to his elder brother comes across in all those little things.

This display of respect to his elders continued long after Dr. Gupta was a renowned statistician and tenured professor, when others were already looking up to him as a mentor and respected professor. His professional stature has never diminished his long-held family values.

Brother, Mentor, and Friend

Imagine a day in May, perhaps 1947, in Purkazi. The mango season is at its peak, the delicious yellow-orange mangos are ripening on the trees, and everything smells like mango. Breathing in the fragrant air, two young boys are walking with Pitaji, their father, to the mango orchard to check on the harvest. When they reach their destination, the orchard manager offers his clasped hands in *namaskar,* a respectful greeting to the entourage, then hands each boy a ripe mango. As Pitaji talks with the farm manager, nine-year-old Arjun and eleven-year-old Sushil head to the shade of the nearest tree to enjoy their sweet, juicy prize. This scene with Arjun's closest brother Sushil Bhaiya was repeated year after year during mango season. As Dr. Gupta shared, "We ate mangos day in and day out. We both loved being in the mango orchard and eating the mangos." Dr. Gupta still loves eating mangos, and he loved visiting the orchards, bringing his family to see them during each trip to India. Describing his unique relationship with his elder brother, he reminisced,

> Sushil Bhaiya was my mentor. We were very close in age—he was only two years older than me—and we enjoyed playing together. As young boys, we loved to go out and fly kites, which was good fun. Part of the fun was that we made kites together at home from colored paper and string. The kites had to be very specialized. We would fly them from the second story of our house. Sometime the other boys would compete to knock other kites down, but we didn't do that.

After the older brothers and sisters moved away—to school, for work, or for marriage—the two school-age brothers Sushil Bhaiya and Arjun, along with their youngest brother, Vinod, had the run of the three-story home in Purkazi for most of the year.

Because the two brothers were close in age, when Sushil Bhaiya blazed the trail to high school in a nearby city and then to Banaras Hindu University, Arjun followed closely in his footsteps. "During high school at Muzaffarnagar, Sushil Bhaiya was ahead of me, so I used his books. Math didn't come easy, but I had the advantage of my older brother's books and notes, and everything else." In addition to his brother's books and notes, Arjun also borrowed from Sushil Bhaiya's experience as he learned to navigate the world outside the sheltered village of Purkazi.

Sushil Bhaiya began to work in the oil industry after college. As part of his work, he went for training in Dallas, Texas, under the Point Four Program, which was set up by President Truman to provide infrastructure training for people from developing countries. At that time, it was not yet common to go to the US from India. Sushil Bhaiya's successful experience overseas also made it easier for Arjun to go to America for his education a few years later.

Sushil Bhaiya and Arjun stayed very close throughout their lives, finding ways to visit each other regardless of where they lived. As chairman of the Oil and Natural Gas Commission (ONGC), Sushil Bhaiya often traveled to the US, making time to see his brother in Ann Arbor and later Bowling Green. When Dr. Gupta and his family visited India, they would always see Sushil Bhaiya and his family, staying with them in Bombay, or Dehradun, or being together in Delhi.

The love between the two siblings extended beyond each other to include both men's families. When it was time for Sushil Bhaiya's son Pankaj to get his master's degree in computer science, Pankaj enrolled at Bowling Green State University, where Dr. Gupta was a professor. Pankaj lived with his aunt and uncle for a short time until he got settled into his own apartment. Later, at

Pankaj's wedding, Dr. Gupta stood in for Pankaj's father, who was in India and unable to travel to New York. Pankaj shared,

> There were so many things that I learned about American life only because of my uncle. I absolutely did not appreciate back then just how easy a lot of things turned out to be for me. I learned from Chachaji how to do things the right way. I would mow the lawn every now and then, which I had never done in India. Even the small things, like going to the hardware store to pick up some screws, nuts, bolts, or washers, and coming back to work on minor projects around the house. It was an introduction to life that was very different from what life was like in India, and I made the transition to American life fairly quickly and easily because of my Chachaji.

Pankaj's ability to adapt to American life in 1989 with competent guidance from his uncle stands in stark contrast to the experience that Arjun experienced earlier in 1963, when he arrived in America for the first time and had to learn everything on his own. Pankaj poignantly described the relationship between his father and his uncle:

> Chachaji cares very deeply about family. There is a lot of respect from Chachaji for his older brother, and they had a lot of love for each other. My dad passed away close to two years ago. During my dad's last days, he thought about his brother constantly and mentioned him all the time. In his last days, when he would call out to my brother, he would actually say, "Arjun," thinking that my brother was his brother Arjun. They felt very tight and very, very close to each other. It was heartrending when my father passed.

An Older Sister

Rounding out Arjun Gupta's birth family are Sarla Prakash, his elder sister by four years, and Vinod Kumar, eight years younger than Arjun. Sarla, or *Jiji*, as she was called, was close in age to both Arjun and Sushil Bhaiya, so the three of them played together often when they were young. Jiji was thrilled to have another baby brother and doted on baby Arjun as he grew into a toddler. For Arjun, having an elder sister at home made the celebration of *Raksha Bandhan* especially joyous. On Raksha Bandhan, a traditional Hindu holiday, sisters tie a string, or *rakhi*, around the wrists of their brothers, offering them sweets and doing *puja* for their brothers. In turn, the brother promises to take care of his sister, offering gifts and money. The celebration gave Arjun a tangible way to express his love and reverence to his sister. Even when he was away at school, he would find a way to come home for that holiday. For many years after, even while Arjun was abroad, both sisters continued to send him a rakhi all the way from India for this occasion. One of his daughters would do the honor of tying it on.

When she was twenty-two-years old, Jiji married and left the Purkazi home, traveling with her husband, Bhoj Prakash, who was a doctor in the army. Jiji and her children moved from location to location with her husband, with one exception. In the 1950s, her husband was posted to Nagaland, which was experiencing an insurrection. Since it was not a family station, Jiji and her children returned to the Gupta ancestral home in Purkazi, renewing her close relationship with Arjun during this stay.

The strong bond of love between Arjun and his sister continued throughout their lives. Jiji's son, Retired Air Commodore Vishwanath, described seeing his mother and uncle's loving interaction.

Arjun *Mamaji* had come to look us up on one of his trips to India. I was very small, maybe six or seven years of age. He walked in, and when my mother became aware that he had arrived, they hugged each other for what seemed like an eternity. My God, just looking at two of them—well, it was quite an experience.

The Youngest Brother

When Arjun was about eight years old, his mother delivered another healthy baby boy, Vinod. Arjun, now no longer the baby of the family, was old enough to enjoy the new addition without any sense of sibling rivalry. Dr. Gupta shared, "He was a wonderful younger brother. He respected me, and we loved each other. He played with his own friends but off and on, we played together." In the same way that Sushil Bhaiya paved the way for Arjun through school, Arjun in turn helped Vinod, sharing books, guidance, and affection.

Following the Gupta tradition, Vinod, too, was given a good education, obtaining a bachelor's and a master's degree in chemistry at Muzaffarnagar. Eventually, Vinod decided to devote his career to managing the farms and helping his parents, since all the other brothers were well-established in their own careers by that time.

Later, when Vinod moved away from Purkazi to nearby (23 km) Roorkee, he opened his home to Arjun and his family as if it was their parents' home. Dr. Gupta's daughter Mita has vivid memories of family visits to Roorkee.

> During almost every trip to India as a family, the home base was in Delhi. Then we would carve out about a week to go to Roorkee and stay with Vinod Chachaji. From there, we would go to Purkazi to spend the day, see the family house, and go to the farms, where we'd sit with the mango trees and eat fresh mangoes and lychees. The town of Roorkee

was along a canal, which came from the waters of the Himalayas, and Vinod Chachaji would arrange trips for us to explore Haridwar and other places in that area.

Vinod passed away when he was only in his early fifties, a tragic, untimely death from leukemia, and a devastating loss for his wife and three children. As one would expect from Dr. Gupta, for whom family is everything, when his younger brother was diagnosed with leukemia, he made every effort to find out treatment options, despite living thousands of miles away. He provided whatever support he could from a distance. He was heartbroken when Vinod ultimately passed away, and it was made even harder being so far away.

This brief narrative paints a picture of the very loving home environment that shaped Arjun Gupta. He had lots of company when the siblings were at home, and they found many simple ways to enjoy their time together.

> My brothers and sisters were a very close family, and we all played together, whatever we did. We had swings that we loved. We had a black dog who lived with us. We called him *Kalu*, which means black in Hindi. We had lots of animals—other dogs, six oxen, and a cat—but Kalu the dog was my favorite.

Looking back on his childhood, and the wealth of rich emotional connections with his siblings, Dr. Gupta shared, "It's nice to have sweet memories of my siblings, but sometimes it's torturing. We don't have each other now. Sometimes I wonder why God Almighty takes everything away from you."

Even though he has become a widely traveled and well-known figure in the world of statistics, Dr. Gupta remains a loving and caring figure in his extended family—a devoted brother to his siblings throughout their life, and a beloved uncle to his nieces and nephews. He has always balanced the lofty realms of abstract statistical theory with his dedication to family, never forsaking one for the other.

~ Chapter Three ~

Knowledge is Power

An unswerving dedication to education is threaded through the Gupta lineage from grandfather to father to sons, and from Arjun Gupta to his children and grandchildren. All of Arjun's siblings encouraged their children to pursue a strong education and succeed. This included educating their daughters as well, which was unlike the archaic Indian custom of only educating boys. For example, when Bhaiya's daughter Jaya, based in Delhi, was pursuing a master's degree in textiles, she went to Bowling Green State University where her uncle, Dr. Gupta, was faculty and stayed with him for a semester.

While Dr. Gupta's life-long commitment to education can be traced back to his grandparents and parents, it completely became his own passion. This commitment is reflected in one of the many sayings he has passed on to his children: "Knowledge is power." One could say that education is synonymous with Arjun Gupta's life. His nephew Pankaj observed, "The one thing that stayed with me is his whole emphasis on the importance and value of education, beyond just the cliché. Chachaji lives that. He cares about it. He talks about it."

As a result, Dr. Gupta expected all three daughters to pursue the best and highest education. Mrs. Gupta shared that when their daughters wanted to pursue their education and interests, Dr. Gupta said, "That's fine. Don't worry about anything. I will provide everything." And he did. Mrs. Gupta continued, "All three daughters went to top universities and whatever they wanted to pursue, they pursued. He was very supportive about their education."

This thread continues forward to the future. "I have three grandchildren—Arhaan, Saamik, and Anika," Dr. Gupta commented. "I hope they get as high an education as possible. With an education they can live anywhere and do anything."

Arjun's Early Education

Arjun was very motivated even as a child to succeed in school. According to Mrs. Gupta, "He was very hard working, and he was always the top. He was always first in the class from his very childhood." Getting an education was not without hardships, yet despite the challenges, he always persevered and worked hard. Seeing his older brothers succeed inspired Arjun to follow in their footsteps, as they were all very close to each other.

The three-storied Gupta family home in Purkazi, with its twelve rooms, made a perfect schoolhouse for the elementary education of Arjun and his younger brother, Vinod. It was like having a private classroom with a favorite classmate. Although his sisters were also educated at home, they had their own teachers and often studied other subjects. Dr. Gupta recalled, "It was fun because the teachers used to come to our home and teach us. We made our own chalkboards, *takhtee*, out of a wooden board. There was also an ink pot that we used for writing." The image of working with these simple tools stands in marked contrast to the later scene of Dr. Gupta teaching his statistics students using Powerpoint slides stored on a thumb-drive, something not even imagined when he was young.

When Arjun was old enough, he went to middle school at Purkazi. He walked or rode his bike through the village to reach school each morning. Returning in the evening, he might stop to get an occasional sweet from the sweet seller. On the fifteen-minute walk down the road to the small village of Purkazi, there was a cacophony of sounds, sights, and smells. Bicycles, rickshaws, and horse-drawn *tongas* weaved their way around each other and around the ever-present buffalo, cows, an occasional chicken, and of course, the pedestrians. Today, very little has changed since Arjun went to school in Purkazi. Perhaps the most visible effect of the modern world is the increased number of cars and motorcycles, which were rare when Arjun was growing up. There is very little in this rural scene that hints at one of its most unexpected products—a world-famous statistician.

The city of Muzaffarnagar became the focus of Arjun's life during his high school years, marking his first transition away from the sheltered life of the village. He was only twelve years old when he entered the four-year high school. The city was sixteen miles away from Purkazi, so he and his brother Sushil Bhaiya, who was already a student there, stayed in a hostel about a half-hour away from the school. The high school was a hub for students from most of the smaller towns, and perhaps for the first time, Arjun mingled with people from many different backgrounds and classes. However, he was well-prepared to mix with others and study since he was fluent in speaking English and Hindi.

Being away from home was not easy for Arjun, despite having his brother there. As Dr. Gupta later shared, "High school was a very difficult time. In high school, for the first time, I was away from home." There were small breaks from study, and unlike the small village of Purkazi, the city of Muzaffarnagar had a movie theater that played the latest Bollywood movies. One afternoon, the two brothers heard that the theater was showing a new movie, *Daag*, with the upcoming young star Dilip Kumar. They took some of their precious spending money and went to see the film. With great feeling, Dilip Kumar sang, *Aye mere dil kahin aur chal*, which translates to, "Oh my heart, let's go somewhere else." Ever since that time, when Dr. Gupta hears the song, he said, "the song brings up memories of a hard life, because I was in high school, away from home, and preparing for my classes."

The antidote to homesickness was returning home regularly. The two brothers went home to Purkazi about every two weeks, and for the many holiday celebrations, such as *Diwali*, the festival of lights, when all the family would gather. Arjun always looked forward to coming back home during the holidays.

> With my three brothers and two sisters, it was quite a fun family gathering. We fought, we played, we ate, we did everything. We ate mostly Indian food, like pan-fried *rotis*. I always ate a special sweet called *gulab jamun*, which I really loved.

The entire house was decorated with festival lights for Diwali, and after dark, everyone went to the terrace to light firecrackers and sparklers.

Onward to College

When Arjun graduated from high school in 1954, he followed his brother Sushil Bhaiya to study at Banaras Hindu University (BHU) in Varanasi, first getting his intermediate college education, followed by the diploma courses. Arjun was only sixteen years old when he entered college. "When I got to Banaras, it was quite hard," Dr. Gupta explained, "because that was the first time I went so far away from home, and especially from my mother."

He could no longer return home every two weeks because of the distance, so the initial transition was difficult. But the presence of his caring siblings once again provided much-needed support for

Arjun's inaugural trip into the wider world. His sister Jiji was living at Allahabad, where her husband was posted, which was driving distance from school. Arjun's visits to his sister provided a loving home environment and also compensated for the mediocre food at the hostel where Arjun lived. Entering Jiji's spacious home, the aroma of home-cooked food immediately eased the pang of homesickness. Both he and his brother Sushil Bhaiya always enjoyed their visits to Jiji and her family. Jiji's son Vishwanath noted that the comfort of the visits was mutual between the siblings. He said his mother often spoke fondly of his uncle's visits during that period, explaining that "she was also very strongly connected to him so it would have been definitely a mutual support."

Having his older brother at Banaras was also a great help. Dr. Gupta said, "I got special attention because my older brother was already there." Young Arjun also was the recipient of some very handsome, up-to-date outfits from his brother, who maintained the family tradition of being well-dressed and had very dapper Western clothes. The two brothers spent almost six years together at BHU, where Arjun was in the Science College while his elder brother was in the College of Engineering.

One could imagine many adventures that the two well-dressed, very handsome young men might have had at the University, if it weren't for the fact that both of them were utterly committed to studying and succeeding. They were quite unified about not wasting their time or their parents' money. While studying was a priority, they did make time for a few treasured excursions to nearby pilgrimage sites, such as Bindhyachal and Sarnath. While reviewing photos from those days, Dr. Gupta mentioned, "Sarnath, where Lord Buddha gave a sermon, is only thirteen miles from Benares, so we could bicycle there."

Other than the occasional outings, Arjun worked extraordinarily hard. In those days, there were frequent power cuts and, of course, no computers, so everything was done by hand. Looking back on his college days, Dr. Gupta noted:

> Life at school was very hard, very difficult. We studied with oil lamps that we had to keep lit. The day was quite long. I got up early in the morning, and sometimes I studied to two or three a.m. and then went to bed. It was really challenging. However, it was not that bad, given our motivation.

His dedicated work clearly paid off, as Arjun became first class, third position in the whole University, and his status opened doors for him, especially in choosing his classes.

In hindsight, Dr. Gupta found his academic education at BHU was quite enriching and broadening, especially in comparison to his later rigorous studies at Poona.

> My brother was there, and I had a chance to take a variety of classes, including French and English. My English teacher, Professor Mukherjee, was fantastic. In addition to statistics, I did a diploma in French, and I also did honors courses. Banaras was very rich for me.

Academic learning was only one facet of Dr. Gupta's college education at Banaras. Another skill that began to emerge in college, and one of the hallmarks of Dr. Gupta's later life, was his remarkable ability to travel anywhere, in any situation, and somehow make it all work. His flexibility and sense of adventure seem to have begun early on, perhaps in his college days.

Arjun always traveled home by train each year in November when *Dassera* and *Diwali* were combined into one month-long holiday. The train ride from Banaras to Purkazi was a long, twenty-four-hour journey. The train always had a first-class compartment, but Arjun took third class, sitting on an uncomfortable wooden bench for the entire time. Dr. Gupta described one particular trip.

> The train itself was quite an experience. One time while traveling, I went to the dining car, and I told my fellow traveler to look after my bags, and he just made away with them. It was very shocking. I pulled the emergency chain and the train stopped. The police took me to the resting room at the train station where I kind of half slept. Then I took the morning train without my bags. In my pocket, I still had some money, so that helped.

Arjun Gupta always had an inner sense of how to cope with things no matter what happened. His ability to deal with daunting situations turns up again and again, particularly during his trip to America and his subsequent travels around the world teaching statistics.

The Nascent Statistician

India is the birthplace of some of the most brilliant mathematical minds, and the field of mathematics is widely respected, making it an excellent profession, along with medicine and engineering, to show oneself. Since his eldest brother was already a doctor, and Sushil Bhaiya was an engineer, Arjun was ready to do something different. Therefore, most significant in Dr. Gupta's career trajectory was the decision to focus on statistics, a decision that came to life while he was at Banaras.

> At Banaras, I was already thinking about statistics. The reason I majored in statistics was because in India, it was the only subject that was really new. My brother, who was

already at Banaras, advised me to pursue it because it was such an upcoming field. So I did.

At BHU, statistics was introduced at the undergraduate level beginning in 1950, four years prior to Arjun's arrival at school. It was not until 1984 that a separate Department of Statistics was created, headed by S. N. Singh. Because of that, Arjun's studies at BHU were in the statistics section of the Mathematics Department in the Science College. Since statistics was a new subject, it was very difficult to find well-trained teachers, and Arjun had to work doubly hard. But working hard, by now, was clearly Arjun's way of life. He shared, "The courses were difficult, but that didn't matter to me."

One semester, all the students in mathematics were abuzz with excitement. Word spread that the pioneering and brilliant statistician, Professor C. R. Rao, had been invited to give a lecture to the students. Professor Rao was only thirty-four years old at the time, but his reputation had already spread across India. Six years earlier, he had received his PhD from Cambridge University in the UK, as the only graduate student of the founder of modern statistics, Sir R. A. Fisher. Now the lecture hall was full, as Arjun and his fellow students waited in anticipation for the young statistician to begin to teach. After the lecture, Arjun politely waited his turn with his classmates for a moment to introduce himself to Professor Rao and exchange a few words.

Even just one meeting with this legendary statistician provided encouragement to Arjun's plan to pursue statistics. Professor Rao went on to become India's living legend of statistics, contributing breakthroughs in the field, including work in multivariate theory, which later became Dr. Gupta's field of specialization. Years later, after Professor Rao moved to the US, Dr. Gupta connected with him more personally, hosting a party for him at his house. "He came to Bowling Green at my invitation to give a talk, and I also introduced him to people. Later, when I was compiling the book in honor of my professor, Dr. Pillai, I asked him if he would submit a paper, which he did." Years later, Dr. Gupta's daughters met with Professor Rao at a Statistical Institute meeting in Tokyo, Japan. In this way, the relationship between these two mathematicians resonated across the world and included their families.

Another major contribution to Dr. Gupta's evolving life as a statistician was a series of almost magical coincidences. One might more accurately say they were the unfolding of Dr. Gupta's inevitable destiny. The head of Arjun's department at Banaras happened to be the brother-in-law of V. S. (Vasant Shankar) Huzurbazar, the head of the Department of Mathematics and Statistics at the University of Poona. Huzurbazar was invited by his brother-in-law to Banaras University for a visit to the Department of Mathematics when Arjun was in his last year as a student there.

Here destiny takes over. Huzurbazar met this aspiring young statistician, who had excellent grades and a top rank in the University. Impressed with Arjun's capability and sincere intention to pursue statistics, Huzurbazar invited Arjun to embark on graduate studies with him at Poona. As Dr. Gupta related in a very droll and drama-free summary, "He said, 'You come with me.' So I went with him." Therefore, directly after graduating from Banaras, Arjun headed to Poona for the next stage in his journey to mastery in statistics.

University of Poona

Going to Poona represented yet another level of self-sufficiency for Arjun Gupta. Unlike Banaras, where he had the company of his brother and nearby sister, he did not know anyone in Poona other than Professor Huzurbazar who had invited him. Arjun's father provided an introductory letter to someone his father knew, but other than that Arjun was quite on his own. In addition, Poona was also a cultural challenge; it is in the state of Maharashtra, where the primary language is Marathi not Hindi, and the food is Maharashtrian rather than North Indian. It was more expensive to go to Poona as well. Dr. Gupta explained, "The hostel was not given to me for housing, because I was not Maharashtrian. I was not a local person. So I stayed in a hotel for a couple of months." In short, as Dr. Gupta shared, "It was very, very difficult." Once again his willingness to work hard and overcome obstacles paved the way. Within a few months, because of his excellent academic standing at the new school, he was given a place in the dormitory, which both saved on expenses and provided some companionship.

Arjun went to Poona specifically to get a master's degree in statistics. However, on the way to the master's degree, he first got a second bachelor's degree, honors, in statistics, then he got the master's degree. Despite all the challenges of language, food, and isolation, he came out with honors, at the top of his class, achieving the first rank.

Arjun's education in statistics expanded in new directions through his study at Poona. Professor Huzurbazar, who was also a former graduate of BHU, was known for his work on sufficient statistics, a field somewhat new to Arjun. In 1974, Vasant Huzurbazar received the Padma Bhushan, one of the highest honors in India, a national award for his contributions to the field of statistics. Dr. Gupta described the academic scene:

> At Poona I found very intelligent students, and the teaching quality there was very good. Vasant Huzurbazar, the department head who had invited me to come, had been trained at Cambridge and then came back to teach at Poona. Since I was there for a master's degree, I didn't work with him as one does for a PhD. But he was in charge of our dorm, the department, and everything else. I took half

my courses from him. He was a very methodical teacher, and he was very strict. Since he was trained at Cambridge, his teaching style was quite different from traditional Indian teachers, in the sense that he challenged the students and asked them to participate instead of just lecturing to them. This teaching style was not common at that time in Indian schools.

While statistics was a very new field of study in the world at that time, India was one of the first countries to embrace it. The University of Poona, for example, had fully established a Department of Statistics, offering degrees in statistics. Arjun Gupta now had the chance to learn from some of the luminaries in the field who visited the school. One time, Sir R. A. Fisher from England was invited by Professor Huzurbazar to give a talk at the University while Arjun was studying there.

> Hearing him speak made me feel that I knew very little. I was already on the track for statistics at that time, but I was just starting, and he and others had taken off.

Dr. Fisher was already seventy when he came to Poona, and he died only two years later, so that encounter was both edifying and precious for Arjun, a young, aspiring statistician.

There was not much time for excursions from Poona, except perhaps to the army club at the nearby training center for the Indian Army, where he could swim and have a nice meal. This was quite different from the pilgrimage trips near Banaras. Again, as he did at Banaras, during the combined long vacations, Arjun traveled home by train, an even longer trip than the Banaras trip, around thirty-eight hours altogether.

While he worked and studied very hard for his education, Arjun Gupta also had (and has) the gift of a brilliant mind, which he brought to all his studies. His nephew Pankaj offers a glimpse of this:

> My dad Sushil always re-told the story of a time when my uncle had a major exam at Poona. Chachaji came out fairly early from the exam room and sat in the hallway waiting for his classmates. It was quite a long while later before all the other students came out. Apparently, that was the first time my uncle discovered that it was actually a very difficult paper. Until then, he didn't realize that this was actually a tough paper. It just went smoothly, and he got it done. We all had such a good laugh out of that. He was really a mathematical genius, according to my dad.

This combination of dedicated work and innate genius brought great results, not just at school, but throughout Dr. Gupta's subsequent career in statistics.

Now the trajectory of Arjun's statistics career took a few unexpected turns. Graduating with honors from University of Poona in 1962, Arjun stepped into his first teaching position. Because of his outstanding grades, he was offered a teaching position at Agra University, where he taught statistics in the Science College. Agra was an excellent choice, as it not only had a notable reputation for quality education, but it brought Arjun closer to home again. Agra is in his home state of Uttar Pradesh, only a five-hour bus and train ride from Purkazi.

However, while teaching at Agra, the unfortunate news arrived that his beloved father passed away. Arjun finished up the semester and then immediately returned to Purkazi to manage the family affairs, until his younger brother Vinod finished his studies and could take over the estate. Once everything was settled, Arjun was offered another teaching position, this time at his former school BHU, where he was invited to teach graduate students.

Again, this was an immensely important move because at that time, Sharadchandra Shankar Shrikhande, a renowned mathematician, was a professor of mathematics at BHU. With some modesty, Dr. Gupta mentioned that "I heard Dr. Shrikhande teach at Banaras when I was a lecturer there. We were very fortunate to listen to him." In this recollection, he humbly omitted the fact that it was Dr. Shrikhande himself who hired Arjun Gupta, which says a great deal about his early recognition in the field. Dr. Shrikhande had received his PhD in the US at the University of North Carolina, in Chapel Hill. He passed away at the age of 102 in 2020.

It was during this time, teaching at BHU, that Arjun made the life-altering decision to pursue a PhD in the United States, launching him on the defining trajectory for his career.

- PHOTO GALLERY -
PART I: INDIA

Baby Arjun at home in Purkazi

Above: Purkazi Road Sign (in Hindi) Left: Village of Purkazi

Part 1: India

Arjun's Grandfather,
Lalita Prasad Gupta (*Baba*)

Arjun's Grandmother,
Manbhi Gupta (*Ma*)

Arjun's mother, Leela Gupta
(*Bhabhi*)

Arjun's father, Amar Nath Gupta (*Pitaji*)
in the family mango fields

Arjun's older sisters, Sarla *(Jiji)* on the left, and Shakuntala Devi *(Bibi)* on the right

Amar Nath Gupta *(Pitaji)* and his eldest son Ram Nath *(Bhaiya)* at Irwin Hospital, Delhi where Ram Nath was living.

Arjun with his older brother Sushil

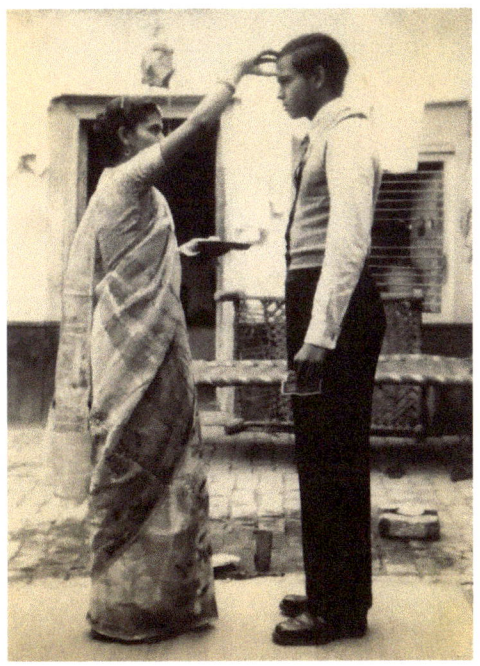

Arjun with elder sister Sarla *(Jiji)* in the courtyard at Purkazi on Raksha Bandhan

Three brothers: Arjun (left), Vinod (center), Sushil (right).

Youngest brother Vinod, Meera Gupta, Arjun Gupta,
in mango fields near Purkazi (Bhujeri)

Dr. Gupta in front of the entrance to the family house in Purkazi

All siblings in 1988: From left to right, Sushil *(Bhaiya)*, Shakuntala *(Bibi)*, Sarla *(Jiji)*, Vinod, Ram Nath *(Bhaiya)*, Arjun

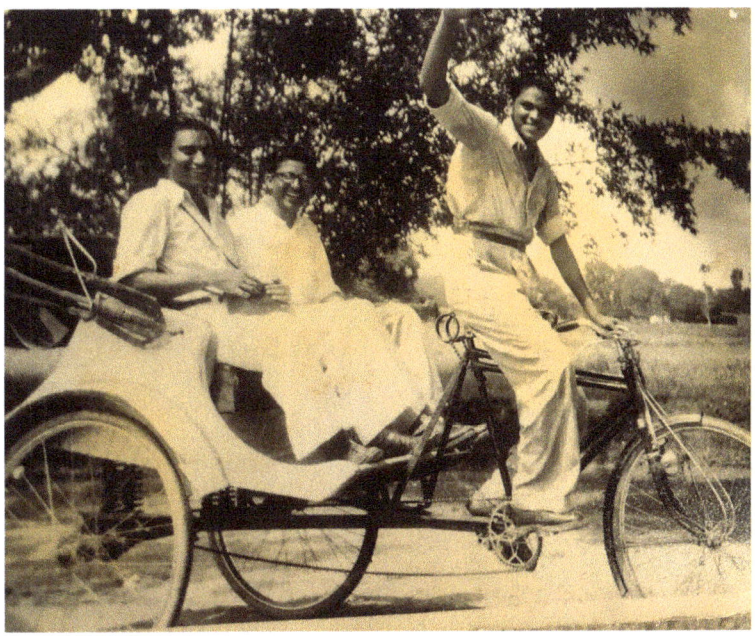

Arjun Gupta (driver) with his older brother Sushil (left) and Sarla's brother-in-law visiting Bindhyachal during college years

Banaras Hindu University, where Arjun Gupta
received his undergraduate degree

At the hostel while attending
Banaras University, 1954

Arjun Gupta, intermediate
graduation from Banaras Hindu
University, 1955

Part II: America – A New Vista Opens

Make courage your companion and you will continue to succeed in everything you do.

Arjun Gupta

~ CHAPTER FOUR ~

IN PURSUIT OF A PHD

Pursuing a PhD in statistics was the obvious next step for Arjun Gupta, but it was not obvious where it would be best to do this. Many of the illustrious Indian mathematicians that Arjun had encountered thus far had gone to the US to study. But going to the US was not an easy decision to make.

> For higher study, I could do it in India or the United States, but in the US, I'd get better opportunities. Talking to my elder brothers, they both encouraged me to go.

> Bhaiya, the eldest, was especially eager for Arjun to go abroad for further studies, as he had a very high opinion of his brother's talent for math and statistics, and he wanted to see it nurtured. Bhaiya did not think India would have a lot of opportunities for Arjun. According to Bhaiya's daughter Jyoti, he was very strong about it, saying to his younger brother Arjun, "You have a lot more prospects going abroad and doing bigger and better things." Bhaiya's voice held a lot of weight since he himself had spent seven years in the United States, Dr. Gupta explained.

> He was the first one from the family who went to the US to study, and he really encouraged me to go to the US when

the time came. It was more common for doctors to go overseas and study, and not so common for mathematics at that time. But statistics was very new, and it was hard to study in India, so I went overseas.

In 1963, when Arjun was ready to go to the US, it was also not that easy to get a student visa. It would be another two years before the Immigration and Nationality Act was signed by President Johnson, removing bias against Asians applying for visas. Although the atmosphere in the US was opening up, it was still considered rare for an Indian student to study in the US. Arjun would be among the first generation of students that came from India to the US and as such, it was a historic step. His nephew Pankaj, who himself came to the US decades later, shared an interesting perspective.

> At that time, the US was like the final frontier as far as we were concerned. It just seems so common now, but then it was extremely uncommon. We try to find Indians in the US who are older than my Chachaji, and they're not there. These guys were among the first to come to the US. They broke through that barrier and made it happen. I don't know how they came to that decision, but it was a big deal to just do it. They didn't have any contacts here. Compared to my arrival, it was like night and day—I had it so easy.

Once the choice was made to go to the United States, Arjun needed to determine which schools would be most suitable. It was important to be selective, since it cost time and money to apply to multiple schools. In addition, statistics was a relatively new field in the US so only a few schools offered PhD programs in it. In the end, Arjun applied to four schools: Berkeley, University of North Carolina at Chapel Hill, Columbia, and Purdue. It took quite a bit of ingenuity to go through the application process from India. There were admission tests to take and applications to mail. The American Embassy and USIS in Delhi each had pieces of the puzzle. However, as is typical for bureaucracies, that meant waiting in long lines outside the embassy just to find the right person with the right papers who was willing to help. Arjun took multiple trips to Delhi, staying overnight with his elder sister Bibi as he waded through the red tape just to apply to graduate programs. This would be repeated when he later applied for his student visa.

After months of anticipation, four thick, white envelopes arrived, one from each school, addressed to Arjun Gupta, announcing his acceptance to their statistics PhD program. It was an unprecedented and exciting moment for Arjun, who now had the luxury of a choice to make. When asked how he made his decision, Dr. Gupta responded, "It's difficult to say how I chose Purdue. But my condition was that I should be supported financially. Purdue gave me best the teaching assistant role with the most financial support." At that time, few schools could match Purdue's well-known mathematics and statistics program. Ultimately, Arjun's choice of Purdue proved to be extremely wise, not just for the quality education he received but also for the connections he made, which proved invaluable for his future career and travels.

For a young man like Arjun, contemplating and deciding to leave India in the 1960s for education in the US involved more than just the mechanics of applying for admission. The significant financial considerations as well as the distance from home also factored into the decision. Dr. Gupta's friend and Purdue classmate Dr. Pradeep Gupta vividly explained his own process.

> Question number one is, can I afford to go to the US? For the student visa, the American embassy insisted on your having access to certain funds in order for them to allow

you to go to America, so we had to figure that out. And as students, we, of course, didn't have anything. It was our family's money. Second, we were going there for many years. Most of us came for a PhD, which was four or five years. The question was, do you want to leave your family for five years? Because there was no chance in hell that you could come back to just visit. So it was a rather difficult decision. I'm sure that same kind of thing must have happened to a lot of students. It was something that they had to work hard to make happen. It was not plain sailing. There may have been some who came on full scholarships but not many of us.

For Arjun, the precedent set by his older brothers as well as his family's emphasis on education made his decision easier.

Leaving the Motherland

In August, 1963, twenty-five-year old Arjun set out on a journey that was an order of magnitude more difficult than going to high school or college, as he was stepping into an unknown culture on his own. As he recalled, "It was very hard to leave home, and I cried at first. But my two elder brothers were really there for me. My eldest brother had gone overseas so he knew what it was like."

In the early 1960s, air travel was considered quite adventurous, rare, and also very expensive. The cheapest ticket from India to the US was on the Russian airline Aeroflot. With an emotional but stoic goodbye to his family, Arjun courageously boarded the Russian plane at Delhi airport for a flight to Moscow. From there, he would fly by way of London to New York, the whole trip taking at least forty-eight hours including the two layovers. The journey to the US was filled with unpredictable twists and turns, calling into action Arjun's capacity to gracefully cope with unusual or difficult circumstances, foreshadowing his later world travels. Describing the first leg of the journey, Dr. Gupta recounted,

> Landing in Moscow was very strange because I couldn't speak their language at all. But I remember they had a machine there. You could speak in English and it would come out in Russian. I was very impressed with that.

After a few hours' layover at the Moscow airport, Arjun boarded a new plane for the flight from Moscow to London, where he had an overnight layover before the flight to New York. He stepped off the plane in London Airport (later renamed Heathrow Airport) and headed through transit customs. The crowd of mostly Europeans was crisscrossing the airport lobby, clearly knowing where they were going. Arjun, a bit travel-weary and perhaps looking somewhat perplexed, strategized where to spend the night. Nearby, an array of signs indicated local hotels with prices that, in comparison to India, seemed exorbitant. It was not uncommon at the airports in India to see families stretched out across the plastic chairs or on the floor, getting a night's sleep, but this did not appear to be the practice in London.

As Arjun wandered through the terminal considering his options, an unfamiliar Indian man approached him, as if by divine intervention, and struck up a conversation. As Dr. Gupta explained, "The man from India was very nice to me. He invited me to his house, so I went there and slept for the night. I don't remember his name, and I never saw him again." The open and trusting nature of their father still shocks his daughters when they hear him tell this story.

The next day, Arjun returned to London Airport to catch the thirteen-hour flight to New York, landing at Idlewild Airport, which only a few months later would be renamed JFK Airport. Once again, Arjun demonstrated his uncanny ability to creatively navigate new situations, putting into practice one of his favorite aphorisms, "Maintain the balance of responding to situations with a cool head and to people with a warm heart." Dr. Gupta narrated,

> I landed in New York in the daytime and had almost no money at all. I had $8 and my airplane ticket to Purdue, so I went to the airline counter, and I traded in my plane ticket to Lafayette, Indiana, for cash. I used that cash to buy a bus ticket instead and found my way from the airport to the bus terminal.

It is somewhat astounding that Arjun had the cleverness of mind to cash in his plane ticket. Yet even more astounding is that the airline agreed! Clearly this was before the era of change fees and non-refundable tickets, and it is also possible that Arjun's warm heart contributed to this unique outcome on the part of the airline agent.

Armed with his bus ticket, the next hurdle was finding the bus station. Arjun spotted other young students and discerned they were also heading to the bus terminal. He traveled in their company, winding his way from JFK Airport to Port Authority Bus station in Midtown Manhattan. There, a slight downturn of fortune happened. The bus station was not in the best of neighborhoods, and Arjun must have looked like an easy target. He was accosted by some young hoodlums.

> When I was walking the last blocks to the bus station, I was pickpocketed. Well, I was not really pickpocketed, but before I reached the bus station, two big kids stopped me and took all my money by force.

Now, other than his bus ticket, he had no money to go to Purdue. Most people might panic and perhaps begin to doubt the whole venture, but once again, Arjun seemed unfazed by it. "There were a couple of students going to Purdue on the bus, so they were able to help me."

At this point, he had been traveling for almost two days, including the layover in London. The final leg of the journey was a twelve-hour bus ride. Arjun climbed aboard the large Greyhound Bus to Indiana, found a window seat, and slept the whole way.

A Purdue Student in the 1960s

It was evening by the time the bus pulled up to the make-shift bus terminal in Lafayette, Indiana, the small city of about 42,000 people that was home to Purdue University. As Arjun stepped off the bus, a young man approached him, explaining he was sent by the University to take Arjun to the student office. At the office, Arjun was welcomed, assigned a room in the student dormitory, and given a map of the campus. By now, the arrival in the US was somewhat a blur of new sights, sounds, and smells. Arjun found his way to his dorm room, got a light bite to eat, unpacked a few things, and had a full night's sleep in a bed for the first time in a few days.

The next morning, he was startled by his unshaven face reflected in the mirror, the residual effects of two days of travel. Arjun made some inquiries and located a nearby barber shop where, alas, he had a rude awakening. "I made a mistake the first morning because I went to get a shave, and it cost me $1." At that time, a shave was only 10 shillings in India, or the equivalent of less than one cent in the US. Given how precious every dollar would be for the next four years, spending that much on a shave was a shock he never forgot.

Arjun Gupta's arrival in the US as a serious student was against a backdrop of current events that would shape the world for generations to come. In August of 1963, the US was still struggling with the Vietnam war, and John F. Kennedy had two more months to lead the country before his tragic death. The message from Dr. Martin Luther King on August 28 of that year proclaimed his dream of equality and echoed sentiments borrowed from the great Mahatma Gandhi. In addition, the emergence of the hippie movement in California brought a new interest in everything Indian, from chai to meditation to the sitar that the Beatles adopted. This was the America that awaited Arjun Gupta when he stepped off the bus in Indiana. Pradeep Gupta, his Purdue classmate, described the scene as it evolved later in the '60s.

> It was a time when the US was going through tumultuous times. America was about to pivot from 1960s to the culture of 1970s. Within that change, there was a dichotomy. While the East Coast and the West Coast were in turmoil in post-race riots of 1968 and the anti-Vietnam War going on at that time, a small, isolated, Midwestern campus with red brick buildings and ivy was kind of protected.

How would this cultural environment impact Arjun, who came seeking knowledge and was ready to work hard? In fact, the backdrop became somewhat invisible, as his focus on his studies never wavered. However, for other reasons, Indiana was still a major culture shock for Arjun.

> Living at Purdue was very hard—perhaps up to that time, the hardest days of my life. I was away from home and working very hard. I was also very homesick. My family called from time to time and wrote letters. But it was very difficult to make a call to India. One time I asked to call to Delhi but got connected to some town with some other name. So it was hard to stay in touch with my family. I

didn't go home for the entire four years, until I got my degree.

Arriving in August eased the transition to the climate a bit, but by November, seriously cold temperatures and snow set in, something Arjun had never before experienced.

> I was not used to the cold and snow. When the cold weather came, there was a church that had a program to sell winter jackets for about $2, so I got a jacket. They didn't sell gloves or hats, so I always kept my hands in my pocket. When the first snow came, I was both excited and terrified. Before I left India, I had bought a pair of shoes with leather soles. When I wore them, I slipped on the ice, which was very challenging.

Finances were, of course, another issue and concern for almost all of the Indian students on campus. Dr. Gupta's friend Pradeep explained,

> We all were pretty much self-supporting, because most of us came without any financial backing. We just had enough money to get plane fare to come to the US. As graduate students, we were hired by the department to help the professor as teaching assistants, or if we were doing a PhD, we'd get a research assistantship. That work used to be our source of money, out of which we basically created our whole life—living and all the expenses, etcetera.

With his teaching assistantship, Arjun managed to cover his expenses, spending the first two semesters grading papers and subsequently acting as a teaching assistant to the professors.

Last, but not least, adapting to the food was another major transition, which certainly contributed to being homesick. There was little chance to enjoy delicious, home-cooked Indian food. In addition, Arjun was a vegetarian, which, in contrast to today's host of dietary preferences, was uncommon. Being creative, he always managed to find something "like mashed potatoes, and peas," he said. He described his first encounters with the self-service cafeteria at the University, where he went through the line selecting steamed vegetables like cauliflower, carrots, and potatoes, fully expecting to find *masalas*, or spices, to put on top, only to discover that salt and pepper were his only choice for seasoning vegetables in Indiana.

Things changed a bit during his second year. Arjun was carrying his suitcase to his new room in the graduate dorm when one of his new campus friends, R. N. Sharma, a computer engineer from the IT College, came by.

"Wait," he said to Arjun, stopping him in his tracks. "There's space in the house where I live."

Sharma grabbed the suitcase and led Arjun to a private house, which had four or five rooms for students. Sharma was already living in one room, and Arjun immediately rented another.

> I met Sharma because he came from India, and he faced the same problems as I did. He was also from Uttar Pradesh, from Kanpur, so we had a lot in common. He was a vegetarian and so was I. We both could be vegetarian in a house together. It made life a little bit easier.

At the house, Arjun could now cook his own food. But cooking was something new for him. At home in India, he had rarely been in the kitchen, since the meals were always prepared by others. Ever resourceful, Arjun learned to cook on the spot. He said, with typical understatement, "I managed to eat and make some Indian food for myself." In fact, he cooked quite well, and others enjoyed his food. From time to time, Arjun would visit his new friends, such as Aridaman Jain, Aditya Mohan, and others in the graduate dorm, and prepare a meal for them. Sharing food, especially Indian food, was a big part of making life pleasant in Indiana. Later, Aridaman Jain's home became another pleasant part of Arjun's life at Purdue. Dr. Jain shared,

> A year later after joining Purdue, my wife joined me. Then Arjun used to come over often to dinner at our place. Anjali, my daughter, was born after I had been there three years, in 1966. When she was born, Arjun came to see her in the hospital and she was small. It was not an early delivery or anything, but she was five pounds. So he said, "She's pinky." And that became her nickname. Even now, we call her Pinki, but her name is Anjali.

The Jains and the Guptas eventually became lifelong family friends. When Arjun got married, Aridaman and his wife would be the first people Mrs. Gupta would meet. The children of both families also became friends and still visit each other all over the globe. For example, Aridaman's son, Arvind, visited Nisha in Bangkok, and Mrs. Gupta and Mita visited Arvind in Kuala Lumpur.

Occasionally, Arjun's visits to his host family provided a break from cooking for himself. Purdue had a program matching up local host families with international students to help ease their transition into American life and provide a mini cultural exchange. Arjun was "adopted" by a farmer and his family after his first year, and they invited Arjun to their house occasionally, offering him their home-cooked American food, which might or might not be vegetarian. As a proper guest, Arjun ate whatever was offered. However, they also were curious about his tastes. Dr. Gupta recollected that "when I came to their house, they sometimes invited me to cook some food because they wanted to know what I cooked." But the farmer's wife did not stock any of the Indian necessities like cumin or cilantro, so instead, "I just cooked an omelet with peas in it." Arjun stayed in touch with the family for many years, and they visited him later when he was living in Bowling Green.

The Infamous Impala

Cooking for himself and others meant that Arjun needed a way to get groceries. Despite the stereotypical image of rather flat land in the Midwest, the area around Purdue has a lot of hills. The closest grocery store was "quite far away, down a hill somewhere," so it was a long walk into town and back, carrying everything. In his second or third year at Purdue, Arjun bought a used car for six hundred dollars, an Impala convertible. Then he had to learn to drive, and do so on the opposite side of the road compared to India. He enrolled in a Driver's Ed program at the University and soon had his US driver's license.

Now Arjun could shop for himself and also bring friends along. This, according to his friend Pradeep, made Arjun quite popular. Most of the other students could not afford a car, yet they all needed to get groceries. The car made a big impression on everyone, especially since it was such a contrast with Arjun's dignified and serious outer demeanor. Pradeep recalled it with great animation.

> And think of what car Arjun had? A 1958 Chevrolet Impala, gray-colored convertible—gray outside, red inside. I can still see him sitting and driving around the campus in that gray convertible Impala with the top open, and his right hand on the seat back, driving with his left hand. He was the king of the road. It was lots of fun.

Years later, Dr. Gupta continued to buy successive models of the same car, and the Impala became part of the Gupta family folklore. The folklore spread all the way back to India, as evidenced by a humorous anecdote from Dr. Gupta's nephew Pankaj. "When I was living in India, my dad *(Arjun's brother)* would tell me that the best car in the world was an Impala. So, I knew it as just, like, the best car in the world. When I came to the US in 1988, I was so impressed that my Chachaji had an Impala," Pankaj shared, laughingly pointing out that he never made the connection that his dad's assessment of the Impala was totally based on his brother Arjun's experience with it.

Unfortunately, in his last year at Purdue, Arjun totaled the car. He explained, "I took a turn, a very sharp turn, and the car turned over. Fortunately, I was not injured at all. I didn't even have to go to the hospital." This was a moment when having a good local host family turned out to be quite helpful, as they came to his assistance when he totaled the car. That should have been the end of that car, but it was repaired. The repairs cost seven hundred dollars, and then he continued using it for a while. That car is long-since gone, but if Arjun still had the Impala today, a collector's item, it would sell for over $100,000!

Bringing India to Indiana

When Arjun arrived at Purdue, there were only about a hundred and twenty or so Indian students in total, and they more or less all knew each other. Pradeep described how they created the Indian Students Association, a small social group for Sunday night social gatherings with a $20 yearly membership fee. Every other Sunday, the group met in the basement of the local Catholic Church, which offered the space free of cost. "In order to make it even more attractive, we used to rent Indian movies," Pradeep explained.

We rented the movie for $150, which came on a big reel shipped by UPS trucks from Chicago. It was like a 10-, 20-, or 30-year-old movie. We paid a little bit to a theater to project it, and we filled the theater with those hundred and twenty students plus their spouses, if they were married. Arjun always used to come to those functions.

While this detail about renting movies might seem irrelevant in the trajectory of Dr. Gupta's statistics career, in fact, it is quite meaningful. At that time, everyone in India was in love with the Indian film songs (and perhaps they still are). Throughout the villages and cities of India, radios would be blaring the songs constantly. Then, as Pradeep described, "We went to America, and we were completely missing that. There were no Indian radio stations, nothing. It was just America." It was as if the music stopped. It was a big cultural loss for young Indian students so far from home.

One solution was to rent the movies. An even better solution was having tapes of the Hindi or Indian songs to play at the student house. As Dr. Gupta recalled, "Music and movies brought us together."

Once again, Arjun became quite popular with his friends. One of Arjun Gupta's lifelong interests is the music from Bollywood movies. He grew up knowing all the words and melodies of the songs, and he can still sing along with many of them. He is, according to his friends and family, quite the collector and connoisseur of Indian songs. He had (and has) a huge collection of songs in various formats including tape, vinyl records, and so forth. At Purdue, friends would visit Arjun's apartment to make a copy of some tape, which in itself was not a small thing. In those days, the copying was done reel to reel, and one reel used to run for hours and hours. The technology was such that you needed three large tape recorders: one to play the songs, one to record the song, and one to control the whole process. It was quite the scene. Tapes were often copied and shared from student to student. Pradeep recalled that the tapes were recorded so many times that the quality was very, very poor, and you could hardly discern which song was playing. But it was good enough.

Many of the Bollywood lyrics are filled with emotion, so Dr. Gupta's love of Indian music is yet another unexpected dimension of his personality. The stereotype of serious statisticians is that they live only in the mind rather than the heart. But Dr. Gupta has always lived in both his heart and mind, as evidenced by his favorite song, the very poignant *"Aye mere dil kahin aur chal,"* "Oh my heart, let's go somewhere else," which he first heard in high school.

In addition to Bollywood music, Dr. Gupta also loves poetry, especially the Urdu poetry called *ghazals*, and even at Purdue, he had a large collection of them.

> My brother Sushil had sent me ghazals on a tape, one of the big six-inch round tapes. It would hold about thirty or forty songs. He sent them so that I wouldn't feel homesick so far away. I also had some other tapes with Indian songs, which were sung in Urdu.

Many of the ghazals are about love, and some of them also were used in the old Bollywood movies.

Considering how hard Arjun and his friends studied and worked, having a musical interlude was a healing balm, like medicine for their spirits. Pradeep recalled.

> Sometimes, we just stopped chatting. Then only the ghazals were playing in the background, and we'd be just listening for a long time without really talking to each other. I remember and appreciate those times with Arjun."

Studying Statistics at Purdue University

For the next four years, Arjun diligently worked on his PhD at Purdue without interruption. Those four years were extremely rich for his expanded education in statistics as well as for his education in American life. When he started at Purdue in September of 1963, his cohort in statistics had about fifteen students, five of whom were from India—three women and two men, including himself and his soon-to-be lifelong friend, Aridaman Jain. "The two girls there from Bombay were very smart," Dr. Gupta added. The cohort of students was quite international, which proved to be extremely helpful later in Dr. Gupta's travels. For example, one of his new classmates, Sabri Al-Ani, was from Iraq, which provided the impetus for a later trip to Iraq. "But my main focus was on study," Dr. Gupta shared, "so I didn't have much time for socializing with friends." He went there to learn statistics, and he did. The faculty was also international, coming from India, Belgium, and elsewhere, and he found "the professors were very tough."

In 1963, the Department of Statistics had just been formed as a separate department, although sharing its budget with the Department of Mathematics. This was one of the first pure statistics departments in all of the American universities, and members of the faculty are now well-known across the world. Professor Irving W. Burr was an expert in quality control and headed the Department of Statistics when Arjun arrived. The chair then passed to Professor Shanti S. Gupta, an expert in theoretical statistics, a position he held for the next thirty-two years. Other faculty members later became family friends of Dr. Gupta, including the noted mathematician Shreeram Shankar Abhyankar. In addition, the opportunity to interact once again with the giant of statistics, C. S. Rao, occurred when Professor Rao came to Purdue to give lectures while Arjun Gupta was there

In short, stepping into Purdue was stepping into an extraordinary environment for the study of statistics. Dr. Gupta commented, "The faculty at Purdue was very inspiring, and I give credit to them for my education."

Professor Pillai

Of all the faculty at Purdue, the most significant person for Arjun was his advisor, Professor K. C. S. Pillai. Dr. Gupta explained,

> A professor gets to choose the student with whom he or she wants to work. He selected me as his PhD student, but only after I had gone through the qualifying exams. I worked with him for two years in my dissertation on multivariate tests.

Professor Pillai had joined the Purdue faculty only a year before Arjun arrived. Professor Pillai was born and educated in Kerala, India, and came to the US in 1951 to pursue his PhD at the University of North Carolina. He later became a Senior Statistical Advisor for the United Nations, where his work included establishing the Statistical Center at the University of the Philippines.

Professor Pillai's work at the UN and in Manila were of great interest to Arjun, who later also traveled on behalf of the UN. Arjun became Professor Pillai's research assistant, working closely with him right from the start. Professor Pillai's research was in the field of multivariate theory, and he was the only one in the whole department who knew this theory. Because of his influence, this ultimately became Arjun's field of research as well.

Professor Pillai and his wife were very kind to Arjun. His wife was from Kerala and not as fluent in English or Hindi, but she graciously hosted various students from India. Since Professor Pillai had also moved from India to America, he knew the challenges of adapting to life in America without the support of family and home. His support, both academically and socially, was very nourishing to Arjun, giving him a sense of home. Ultimately, their relationship evolved into a personal friendship.

> It was really fantastic. He invited me to his house for meals. He had two sons and one daughter, so when I went to his house, I was with the whole family. In the University, I had a little office, and he even dropped in my office when I was working, just to come and say hello.

Dr. Gupta's friend Aridaman Jain also offered some insights about Professor Pillai's personality:

> He was very sociable. He had a lot of parties, after seminars and otherwise. Sometimes he would crack some jokes. One of the jokes he used to tell was about India. People here in the US used to make jokes about cows running loose in India. So he would say, "Well, that's the check for the brakes on the car. You know the brakes are working if you can stop when the cow is coming."

Describing Professor Pillai, Dr. Gupta said, "His most interesting characteristic was that he worked really hard. But he didn't have to, thank God. Doing statistical research was the most important thing in his life." Dr. Gupta has also been described as someone who is sociable, worked hard, and enjoyed jokes, which makes his relationship to Professor Pillai even more significant. Working with Professor Pillai, Arjun had a mirror in which to see the best of himself.

It is safe to say that the defining influence at Purdue was Arjun's relationship with Professor Pillai.

Professor Pillai certainly was a real force for my entire stay in Purdue. The most important advice he gave me was "Never give up. Keep working. Never quit, never quit." There were times when I thought this whole thing was too long and too much, and maybe the light at the end of the day was not completely rewarding, and he encouraged me. I ended up writing a paper with him on the multivariate test, which was published in a journal.

It is unusual to have one's paper published before receiving a PhD, so this was a huge encouragement for Arjun early in his career.

Professor Pillai not only expanded Arjun's understanding of statistics, he also became a role model for teaching statistics. Professor Pillai was well-liked by his PhD students, all of whom went on to achieve great success in statistics, highlighting Professor Pillai's skill as a teacher. Dr. Gupta said, "He spent time to make sure that they understood what he was talking about." This influenced Dr. Gupta's teaching later on, and he is well-known for spending significant time with his students. As he shared, "I made sure that they understood what I said, because multivariate statistics might take effort to get hold of."

Upon leaving Purdue, Arjun and Professor Pillai maintained an ongoing relationship. They wrote to each other and met regularly at events such as the Statistical Institute meeting. Professor Pillai was a welcome guest at the Gupta home in Bowling Green, where the entire Gupta family got to know this exceptional man.

Professor Pillai died in 1985, almost twenty years after Arjun graduated from Purdue. As a way of honoring his professor, Dr. Gupta compiled *Advances in Multivariate Statistical Analysis,* published in 1987. He reached out to many renowned statisticians, included C. R. Rao, inviting them to contribute to the collection. In the foreword, dedicated to Professor Pillai, Dr. Gupta wrote:

The death of Professor K. C. Sreedharan Pillai on June 5, 1985 was a heavy loss to many statisticians all around the world. This volume is dedicated to his memory in recognition of his many contributions in multivariate statistical analysis.

Graduating from Purdue

For Arjun, the four years at Purdue were filled with a one-pointed commitment to earn his degree, no matter what it took. While all the international students came with that intention, Arjun stood out for his focus. His classmate Pradeep shared:

> He was a kind of serious student, and he was very dignified. He didn't crack silly jokes. He did not get into all these big chats about campus gossip and that sort of thing. We used to look at him as somebody we would not invite to a Saturday night party, but we would certainly like to go and talk to him.
>
> One of the things I noticed was that he was focused, more focused than many of us. Many of us were curious about America—American culture, American food, American way of doing things. But Arjun was more focused on his goal of being excellent in math and statistics. He has proven himself to be somebody who one could be proud of. His kids should be very, very proud of him for all he has achieved, because we grew up when things were not easy.

Later, Dr. Gupta reflected on the challenges of being at school and credited his father for the ability to stay with it.

Even though it was hard, of course I would not give up. I learned discipline from my father especially. He said that in the world, there are only two things that will serve you. One is how healthy you are, and the second is education. I took that to heart, and later, I tried my best—as much as they care to listen—to spread that to my children. I didn't have any regrets, because that's the life I had chosen, so I had to pursue it.

Arjun was on track for graduation in 1967, along with his other classmates. He attended the convocation at Purdue, but his degree was officially awarded in January 1968. This was because once again, Arjun made a thoughtful and carefully planned decision. He had already planned to travel home to India in December of 1967, where a marriage was being arranged for him. But he knew that if he graduated, he would lose his student visa. So he delayed doing his orals:

I had finished all the work, but I did not have the degree, because I was scared that they would have stopped me at the airport to ask "Why are you going there?" I could say I was returning to finish my degree.

His wife Meera added,

If he had taken the orals, the visa status of the student would have been different. Then I would have had a very difficult time, and I would have had to wait back in India to get in. So he postponed the degree by two months. But his degree became one year older, because had he done it in December, it would have been 1967, and we needed to extend it to '68. So he always tells me, "Because of you, I got my degree a year later."

In December, Arjun went back to India, returning with his wife in January of 1968. His student days successfully completed, a new phase in his life, including marriage, family, teaching, research, and world travel, was about to begin. He would live a life mirrored in the Purdue motto, "Education, Research, Service."

~ CHAPTER FIVE ~

Post-Graduation: New Family, New Locations

One of the guiding forces for Arjun Gupta is his commitment to family, which was firmly established from his early life in Purkazi. Once he completed his studies, he was prepared to widen the family circle by getting married and adding a family of his own to the Gupta lineage. Prior to attending Purdue, he already knew the importance of getting married. Before his father passed away, he told Arjun, "You have to get married first so that your younger brother can get married." Dr. Gupta explained, "My elder brothers were already married and so were the older sisters. I was the only one who wasn't married as well as Vinod, so it was my turn."

According to the customs of the day, an Indian family would typically arrange a suitable marriage by considering family status as well as cultural and religious traditions. While the arrangements could become a long, drawn-out affair, everything was simplified for Arjun by the support of his eldest brother and sister. The niece of Bhaiya's wife (her brother's daughter), Meera Nath, had grown to be a lovely, highly educated young woman by the time Arjun was ready to be married. Bhaiya thought Meera would be a wonderful person for Arjun to marry. Meera (Nath) Gupta describes it like this:

My aunt Sushma, my father's sister, is married to Arjun's eldest brother. Arjun had just finished his education in the US. Then it was time to get married. And his eldest brother, who happens to be my uncle, said, "Arjun, you have come to India. You have to get married." That's it. So Arjun said, "OK, whatever you decide is fine with me." Bhaiya probably mentioned my name. "OK, fine," Arjun said. He was a good-looking guy. But I wasn't bad either!

Arjun's approach to life had thus far always included respecting the guidance of his elders, which had the wonderful effect of simplifying decision-making. As he shared about the marriage, "I didn't have too many thoughts about it. I did whatever my elders told me. My parents were gone by then, so my elder brothers and sisters arranged it."

Meera and Arjun had met each other briefly ten years earlier at Bhaiya's wedding. Meera would have been about thirteen or fourteen, and Arjun, almost twenty, was still in college at Banaras. While he might not have paid much attention to a cute teenage girl, Meera remembered meeting him. He was already at that time a striking figure—tall, handsome, and thoughtful. According to Meera, at one moment during the family gathering, she made chai and served it to Arjun. But in all the excitement of meeting everyone, young Meera forgot to put in sugar, so he didn't drink it. That moment clearly did not stand in the way of their future life together.

In December of 1967, after being away from India for four long years, Arjun made the arduous journey back home to fulfill his commitment to his family by getting married. He traveled by way of Hong Kong and Thailand to Delhi, where he landed during a heavy rain and made his way to the home of his eldest sister. The reunion with his siblings was joyous and emotional. Arjun became reacquainted with his nieces and nephews, who had grown considerably in the four years, and he once again enjoyed the delicious home-cooked food generously prepared by his sister. At this time, Bhaiya, Bibi, and Jiji all lived relatively near each other in Delhi, so visiting everyone was easy. The other siblings, Vinod and Sushil, also came to visit and to witness the wedding.

Two days after his arrival in Delhi, Arjun stepped into Meera's parents' house on Humayun Road to meet his future wife. The social atmosphere created by Arjun's eldest brother and Meera's family was a buffer for any unease that either Meera or Arjun may have felt. Following their initial meeting, Meera's parents hosted an engagement party for the immediate family. The actual marriage took place on December 25. At the festivities, platter after platter of food was offered as the two extended families were introduced. Gifts were exchanged, and everyone spoke encouragingly to the young couple. Many of the invited guests were either meeting Arjun for the first time or seeing him again after his long stay in the US.

The day after the wedding, Arjun and his new bride traveled to Purkazi, along with Arjun's sister Jiji and her children. The arrival in Purkazi was a poignant moment, as much had changed in the four years since Arjun had been there. With his parents gone and his brother Vinod living in Roorkee, there was now a caretaker in the house that had previously been filled with the life and sounds of the Gupta family. But the beauty of the mango orchards remained unchanged. Standing in their shade, Arjun visually took in the maturing trees, imagining the sweetness that would erupt with the ripened mangos in May.

Introducing Meera to the revered family home and beloved mango orchards felt like the culmination of the wedding ceremony, as she was marrying into this rich history of Purkazi. This visit to the orchards marked the first of many such visits Arjun and Meera would later make as their own family grew.

Only five days after the wedding, on December 30, it was time to return to the US. Time had flown by much too quickly for Arjun, and he made a mental note not to let four years go by between visits to his family in India. In fact, he later returned to visit every two years whenever possible.

The scene at the airport was like an extension of the wedding celebrations, with family and friends gathering there to see the newlyweds off. Many of the well-wishers said their goodbyes outside, but Meera's parents, Bhaiya and his wife, and some of the nieces and nephews came into the Air India terminal, and even went out to the tarmac. The newlywed couple was garlanded with fragrant *champa* and jasmine flowers, said their tearful farewells, then boarded the Air India Boeing 707 for their trip to the US.

Seated on the plane next to her husband, Meera Gupta pondered her future. She was not only leaving her familiar way of life but was also going to be separated from her family and friends for who knew how long. However, she was heading to America, which seemed like a dream at that time. Like Arjun, she trusted that all would be well, and she relaxed into the long flight.

The flight from India headed to New York by way of Frankfurt and Paris. Landing in Frankfurt, Germany, for a layover, Meera and Arjun finally had their first solo "date," enjoying a meal together and wandering in the shops of the airport. Up until that moment, they had been surrounded by family night and day. Now they could finally get to know each other and begin their lifelong companionship marked by mutual respect and genuine love. While it might be hard for Westerners to understand the value of an arranged marriage, one might colloquially say, "The proof is in the pudding." With their mutual commitment to make it work, Meera and Arjun's marriage has lasted for more than five decades, withstanding the unpredictable curves and twists of life, and becoming a strong foundation for the Gupta children and grandchildren to build their lives upon. In short, it surpasses most of the statistics for so-called success of romantic marriages!

Landing in New York involved the typical lines and delays of immigration and customs. However, Arjun's student visa worked like a charm. Arjun and Meera were admitted, their passports duly stamped, and the immigration officer recited his usual mantra, "Welcome to America." Outside the customs area, Arjun's good friends from Purdue, Aridaman Jain and his wife Nirmal, were waiting to greet the young couple and drive them to the Jain's apartment nearby in New Jersey, where the newlyweds could stay for a few days.

Now Arjun and Meera had a chance to be tourists in New York City for the first time. As Dr. Gupta observed, "New York is not a town, New York is a country. There are people from all parts of the world. But on top of that, there are characters of all kinds." With the Jain's home as a base, Arjun and Meera took in all the sights, including the Statue of Liberty, United Nations, Times Square, and more. "It certainly was very enjoyable," Dr. Gupta commented.

From New York, they continued the journey to Lafayette, Indiana, where Arjun had to take the orals in order to formally get his degree. They needed a place to live in Indiana, and fortunately, a dear family friend, Ram Avatar Gupta, offered them a place in his home. The kindness of friends and family were often the saving grace for Arjun throughout his life. He in turn offered that support to others as well, so the circle of generosity was never broken.

Returning to school after an absence, studying for orals, being newly married, and preparing to start his first job created some amount of stress. The stress and a combination of factors, including perhaps too much caffeine to help him study and not enough water, led to the unfortunate and painful experience of severe stomach pain, and Arjun landed in the hospital to treat kidney stones. Nevertheless, he persisted, and by the beginning of January 1968, he successfully completed the orals and was ready to head off to his first teaching job.

The New Professor in Arizona

For a new graduate in a relatively new field, obtaining a suitable position required some persistence and ingenuity. Dr. Gupta narrated how he landed at the University of Arizona.

> When I graduated, I was looking for anything that paid me well, and I chose to teach. Spelman College in Georgia offered me a position. I wasn't sure what to say so I told them I would ask my professor. I asked Professor Pillai, and he recommended that I not go as they didn't have a strong graduate program. So I turned them down. Another offer came from the University of Arizona, and that's where I joined.

Arjun knew little about the culture in Arizona, except perhaps that the climate was the polar opposite of Indiana, and akin to warmer regions of India. However, it would certainly not have a large Indian population, as the number of Indian immigrants to America was still low, and those that came headed to larger cities. Armed with the trust that things would work out and his willingness to try new things, Arjun made plans to move there.

Arjun's newest acquisition, a used Chevy Bel Air that replaced his trusty Impala, became the chariot that took Arjun and Meera to Tucson, Arizona by way of Dallas, Texas. The first part of the journey, from Lafayette to Dallas, was a grueling thirteen-hour trip through rather boring landscapes, but Arjun and Meera did their best to make the trip interesting. Dr. Gupta explained,

I went south on the road, almost to the southern border, because I wanted to see Dallas, the place where President Kennedy was shot. We saw the Book Depot where Lee Harvey Oswald was when he shot the president. I had been in school at Purdue when that happened in 1963, so I had the experience that everyone in the United States had, and it was very fresh.

The remainder of the trip, from Dallas to Tucson, was another thirteen-hour drive through dry desert country and unfamiliar culture, heavily influenced by the neighboring Mexican culture. The good news was that they left a cold, snowy, winter climate behind as they headed to a dry, summer-like climate in Arizona. They traveled light, carrying just their clothing, books, and notes.

Arriving in Tucson, Arjun and Meera found a convenient motel for the first few nights and drove around during the day looking for a house to rent. "Ninety-four dollars a month," Dr. Gupta explained, "for a one-bedroom near the campus. Then I used my car to go back and forth to campus." The school session started within days of their arrival, so there was not much time to get acclimated before Dr. Gupta began his new teaching job.

It's helpful to pause for a moment and reflect on these circumstances: Arjun Gupta had never been to Arizona nor did he know anyone there. He had lived in the US for four years sheltered on the campus of Purdue. Now he drove 1,700 miles across the US through various climates and cultures, then navigated the intricacies of finding a place to live in Tucson, a city he had never visited. He had a new house to furnish, a new job to start, and a new life to build with Meera. Yet he did all that was needed as if it were second nature. His ability to be undaunted by the demands of life is reflected in a motto that he shared with his children: "Make courage your companion and you will continue to succeed in everything you do."

Dr. Gupta's first impression of arriving in Arizona, much of which is a desert, was how dry it was. But the University was like an oasis. "The University was in a very beautiful area," Dr. Gupta shared. "Palm trees lined both sides of the road, and cactus was all around." Although it was a relief to leave behind the cold winters of Indiana, the heat was quite extreme in Arizona.

> What stands out about Arizona was that it was too hot: it was a hundred and ten in the shade. Our apartment had a water tank on top that was supposed to cool it, but it wasn't really cool. There was no air conditioning in the apartment.

Dr. Gupta joined the University as an assistant professor of mathematics and shared an office on campus with one other person. Fortunately, his officemate mostly took courses and worked from home, so he had the place to himself most of the time.

Unlike the state of Arizona in general, the Department of Mathematics at the University was surprisingly international, much like Purdue. The head of the department was a British man, and Dr. Rabindra N. Bhattacharya and Dr. Mohindar Singh Cheema, both from India, were also in the department.

> This was good luck for me. Dr. Bhattacharya was in statistics, my field, and I had good relationships with both of them. Dr. Bhattacharya invited us for dinner. We went to his house, and he came to our house, as did Dr. Cheema.

The two statistics specialists, Dr. Gupta and Dr. Bhattacharya, shared the teaching load as they introduced statistics courses in the newly expanded Department of Mathematics.

Arizona was Dr. Gupta's first teaching job in the US, so naturally, there was some nervousness meeting the students for the first time. Statistics was new at the school and only offered at the graduate level. Although the students were motivated, they did not have much of a background in statistics. As a result, any initial nervousness dissolved quickly as Dr. Gupta realized he was well-qualified to teach this group. However, he had to accommodate the material to meet the students' level, which he instinctually knew how to do. Since there were only a few statistics classes, he also taught some mathematics classes.

Initially, Dr. Gupta focused on teaching and writing papers from his thesis on multivariate testing, so there was not much time for other research. Eventually, he was assigned a teaching assistant, a helper who could do computer analysis for Dr. Gupta's publications. Then the research progressed more rapidly.

During this time, Dr. Gupta began to connect with statisticians from around the world, a practice that grew substantially over the following decades. One easy way to connect to others was through various statistical organizations. He traveled to a conference in Toronto hosted by a statistics organization, and in 1970, he became a Fellow in the Royal Statistical Society of England. He commented,

> Some of our senior statisticians had already been elected, so I thought it would be a good thing to join them. I didn't know anybody personally. Later on, I visited the office of the Royal Society in London.

He ultimately was a fellow or member in at least eight learned societies and mathematical institutes, which also played a role in his later world travels as they offered connections in many far-flung reaches of the world.

Another connection that began in Arizona and continued later in Michigan was Dr. Gupta's work with the U. S. Air Force. The Air Force had come to the math department at the University in search of someone who could teach Air Force cadets introductory statistics. Dr. Gupta was a great fit for that, and he spent time during the summers lecturing at the Air Force base at Fort Huachuca, teaching "Statistical Methods for Engineers." This work also allowed him to create his first set of portable lecture slides, using transparencies that could be easily brought to other locations, foreshadowing the many lectures he later offered as visiting professor to universities around the world. His daughters have vivid recollections of their father carrying those transparencies in a box wherever he went. As technology advanced, Dr. Gupta moved the transparencies to a USB disk, making traveling even easier.

Dr. Gupta's own experience of being far away from his home environment translated into true empathy for how others might feel. This motivated Dr. Gupta to establish the Indian Students Association at the University of Arizona. He had started a similar group in Purdue to create connections for students and faculty from India. As he shared, "Even though there weren't that many Indian students in Arizona, you have to make a start somewhere." His activities are part of his lifelong interest in nurturing the connection to the Indian culture for himself and others.

The atmosphere at the University was more relaxed than at Purdue, especially for Dr. Gupta, as he was now on faculty and no longer a student. This transition from a single student on a tight budget to a married faculty member with an annual salary created some space in Arjun and Meera's life to explore other aspects of life in the US. Always an adventurer, interested in the many faces of humanity, they traveled to Las Vegas, of all the unlikely places, just to see what it was like. As Dr. Gupta later shared, "Las Vegas was fantastic. It was an eye opener, with all the lights. We went into the casinos, but not with any intention of winning, just to enjoy being there."

Good friends and good food play an indispensable role in creating comfort for Dr. Gupta. In Arizona, his and Meera's social life centered around the University, as they made friends with the Indian community and neighbors. Meera joined the University of Arizona as a master's student in public administration—which would be Meera's second master's degree—where she met other students. Their friends included colleagues in the math department as well as Dr. Chandola from the Department of East Asian Studies and his wife.

Food was definitely elevated to a new level now that Arjun was no longer a bachelor student. Meera provided delicious home-cooked meals, preparing familiar Indian food despite the remote location of Arizona.

> At that time, we were not expecting there to be an Indian population, and there were no Indian groceries or anything. There was a shop in San Francisco, and there were two or three more Indian families, so we ordered by mail and just divided it. I made *daal* with local things like split peas, brown lentils, red kidney beans, or garbanzos. And rice, of course, was available, but not basmati rice, just Uncle Ben's rice. Some Indian spices were available in Arizona, so I really did not have a problem making things.

In some ways, Tucson was a little reminiscent of India, especially with the warm climate. In addition, Dr. Gupta said, "I could go to Mexico. I crossed the border and got a haircut there for a dollar." However, cheap haircuts and Uncle Ben's rice were poor substitutes for actually being in India. As Dr. Gupta shared, "There were no trips back to India during this time, and that was the hardest thing for both me and my wife."

The good news traveled quickly to India that in March 1970, two years after arriving in Arizona, Arjun and Meera now had a baby girl. Arjun decided to name the baby after his mother. "My mother's name was Leela, so I named the baby Sheela." The birth certificate was signed for Sheela, and the baby even received a baby spoon gift with "Sheela" engraved on it. A week went by. Then Meera, now recovered from the delivery, spoke up. "This name is just way too old fashioned," she said, and other family members weighed in with agreement. "That sounds rather old," they said on transpacific phone calls. So the baby's name was changed to Alka. As Dr. Gupta shared, "I had no say in that," surrendering as he often did to forces beyond his control.

Without the traditional network of relatives nearby for support, Arjun and Meera had to manage on their own. Arjun took over some of the cooking and running of things while still teaching, and for the first few months, he recounted that it was challenging for both himself and Meera.

Alas, few baby videos remain. The family story about it goes like this: one night, the young family had gone to a friend's house for dinner. Driving off to the dinner party, they noticed a black cat crossing in front of them. Thinking of the superstitious sayings around black cats, they considered turning back, despite not being at all superstitious themselves. Upon returning from the party, they saw a man running away from their apartment complex. Entering their apartment, they saw it had been thoroughly ransacked, and among other things, their camera had been taken. A sad event, but one that simply faded into the tapestry of life in the wild west of America.

From Arizona to Michigan via India

Arjun Gupta was on the faculty at the University of Arizona for about three years. Two papers that he published during that time, "On the Exact Distribution of Wilks' Criterion" and "Noncentral distribution of Wilks' statistic in MANOVA" grew out of Dr. Gupta's dissertation. At the same time, Meera Gupta had also been busy, as she completed her master's degree in public administration, gave birth to Alka, and was now pregnant with their second child.

However, Arizona was too remote from everything. At that time, the largest density of educated Indians that came to the US were in the Midwest and Eastern cities. According to his daughter Alka,

> Dad always said he wanted to be driving distance, maybe within five hours, of as many cities as possible. He always thought Ohio was that, which is where he eventually wound up. He felt Pittsburgh, Detroit, Chicago, Columbus, and Cincinnati all represented different ecosystems of Indian nationals. And Arizona was just too far away from them.

More significantly, University of Arizona had no Department of Statistics, only a Department of Mathematics. While Arizona had been a good place to launch his new teaching career and establish himself in the field of statistics, ultimately Dr. Gupta needed an even more focused environment to pursue research and make significant connections in the world of statistics.

After looking into other schools, the University of Michigan came up as the best choice. It would give Dr. Gupta a major boost in research, editing, writing, and teaching. Michigan was undoubtedly one of the top schools for statistics at the time, and many "firsts" would occur for Dr. Gupta, including the first of his many positions as a journal editor and his first excursion as a visiting faculty member to another country. Along with the expansion of his professional career, his family would triple in Michigan with the birth of two more daughters.

After securing an excellent position at the University of Michigan, Dr. Gupta arranged to move there. The gap between teaching positions was an opportune time for a trip to India. It had been three years since either he or Meera had seen their families, and they now had a new baby to introduce to the family as well. All the details were immediately arranged: a company was booked to drive the car to Detroit, a mover gathered the rest of their belongings, and the plane tickets were arranged for the flight to India. There was a small farewell gathering with their Tucson friends, whom they were sorry to leave.

A friend drove Meera, Arjun, and baby Alka to the airport for the flight to JFK Airport in New York, where they changed for the long, international flight to Bombay. Traveling with baby Alka, who was about eighteen-months old, added new logistics to be considered. While there was not a separate seat on the plane for the baby, the airline provided a bassinet that they could use. Meera Gupta was already six months pregnant with her second child at the time, so Arjun often held the baby when she was resting.

After arriving in Bombay, Arjun, Meera, and baby Alka traveled by train to Delhi. The arrival in Delhi was especially sweet, as all the relatives got to meet the new baby, and of course, see Meera and Arjun as well. Surprisingly, however, this time India was a bit of a culture shock for Arjun. The contrasts between India and the US were more apparent to Arjun, especially now that he was traveling as a family man rather than a free-wheeling student. After living in the US for seven years, he had become accustomed to simple comforts, such as good roads, comfortable beds, predictable traffic patterns, and so forth. Now back in Delhi, India, they traveled everywhere by bus on bumpy roads, which was not only uncomfortable but somewhat heart-stopping due to the unpredictable way people drove. The constant blaring of car horns and the mélange of animals, pedestrians, bicycles, and cars on the road were a far cry from the comparatively tame, small city of Tucson.

However, the discomforts were well overshadowed by the point of the visit, which was to reconnect with all the relatives. Visits with his and Meera's siblings nourished Arjun as only family love can do. He and his family also traveled to beloved Purkazi, which has its own special place in his heart. After savoring their brief month in India, the family headed once again to the Delhi airport, where they embarked on the long return journey from India via Europe and New York to reach Michigan. It was August 1971.

"Buckle your seatbelts in preparation for our landing in Detroit," the pilot said over the airplane loudspeaker. As the plane circled the airport, Arjun peered out the window. An unexpected feeling of familiarity came over him as he saw the landscape of the Midwest rather than the stark Arizona desert. It felt a bit like a homecoming, although perhaps not everything was welcome. Once again, the cold weather would be a constant companion for Arjun, who perhaps never really got used to the harsh Midwest winters. But Arjun had a feeling of anticipation as his new adventure at the University of Michigan, a renowned school in his field, was about to start.

At the Detroit airport, the trusty Bel Air Chevrolet that they had shipped from Arizona was waiting for them, and the Guptas drove it for the short trip to their final destination, Ann Arbor. Arjun and Meera went through the now-familiar routine of staying in a motel and searching for affordable housing for their growing family. It was just before the start of the semester, and housing was harder to find in Ann Arbor, since the students were also searching for places to live. Arjun and Meera drove around town looking at apartments, and eventually found a two-bedroom apartment in Ypsilanti, about a twenty-minute drive from the campus.

In October, only three months after settling in and starting his new job, Arjun and Meera's second daughter, Mita, was born. Again, it was a challenge to find support in a new town, and people in Michigan all seemed to be much busier than Arizona. However, the Guptas had already connected with the local Indian community who were quite helpful, and when Meera went to the hospital, baby Alka stayed with a new family friend, Radha Handiekar.

University of Michigan, Department of Statistics

The University of Michigan had been one of the first two American universities to offer academic programs in mathematical statistics, as early as the 1920s. Dr. Gupta's arrival as an assistant professor in Michigan in 1971 coincided with important milestones for the department. The Department of Statistics at the University of Michigan had been founded only two years earlier in 1969, establishing a new graduate program in statistics. The department awarded its first master's degree in 1971 and the first doctorate in 1972.

While the Department of Mathematics was well-established and large, with fifty or sixty people in it, the Department of Statistics was still rather small, with a total of six faculty members. The chair of the department, Bill Ericson, had been there since 1962. He had received his degree from Harvard University, was quite personable, and had a keen sense for the future direction of the department. On a personal level, Dr. Gupta shared that Bill Ericson was very helpful.

> He invited us to his house for dinner and checked on us. His door was always open for us to come in and tell him anything and everything. Through my entire stay at Michigan, Bill Ericson had the most influence on me.

When Dr. Gupta later moved to Bowling Green, he stayed in touch with Bill, and Bill also subsequently visited him there as well.

Dr. Gupta's students were mostly enrolled in the new PhD program, although he also taught some students who were fulfilling the statistics requirement for the engineering program. He observed, "My class load was two courses, three hours each, for six hours of teaching. I always enjoyed teaching the Multivariate Analysis class since that's what my thesis was on. The students here were smart and very focused."

Michigan was an intense and competitive atmosphere, and there was a lot of pressure to publish papers. As a result, Arjun Gupta had little time for enjoying Ann Arbor and the area.

> I had no time for anything else since there was a lot of pressure from work. But the student union was just across from the department. We used to go down there and have tea and coffee, and that was helpful.

Extracurricular Activities

During his time in Michigan, Dr. Gupta began his lifelong association with the leading mathematical journals as an editor and reviewer of articles. The first such position occurred in 1973 when he became a consultant for *Mathematical Reviews,* a publication of the American Mathematical Society. In addition to further establishing Dr. Gupta's name in the worldwide statistics community, this role turned out to be an excellent source of additional income since he was paid for every review. His contributions to peer-reviewed articles demonstrates his enormous dedication to the field. Each article had to be read, understood in the context of what went before it, and then reviewed in a way that made it accessible to worldwide readers.

I reviewed other people's papers. I had the key to the math department at Angell Hall, so I could go there when I was done with my Department of Statistics work. I could even take those papers back to my own home to review them there.

This innocuous statement also reveals the emergence of another lifelong pattern: bringing work home. His daughters commented that their father was always coming home with a bundle of papers to work on, either student papers or journal articles. In short, there was a very porous boundary between home and work life for Arjun Gupta, yet he remained available if one of his daughters needed his help with homework or wanted to watch an episode of one of their favorite TV shows with him.

Dr. Gupta's role in reviewing statistical journal articles is also an example of offering service to others in the statistics field in a very simple and tangible way. Explaining the purpose of reviews, he said:

There are thousands of people in the country who would like to know what is in a paper and whether it is worth reading. When writing a review of a statistical or a mathematical paper, I'm looking for originality. Originality of the results is the main criteria.

Dr. Gupta was a consultant for *Mathematical Reviews* from 1973 to 1976, during his tenure at Michigan. It was impossible to continue later at Bowling Green, because he had to be there in person to access the articles for review. There was no electronic media at the time, which is quite the contrast to the situation today, where all the articles are accessible via computer, and reviewers can work remotely.

Dr. Gupta was also involved in *Creations in Mathematics*, a journal that published articles in both English and Polish. The editor of that journal, who was from Poland, had come to Michigan for a year and met Dr. Gupta there. When he was asked to join as an editor, Dr. Gupta declined due to his preexisting commitments, but joined the journal's board to help out, sharing later, "I did whatever I could."

The participation in journal reviews started slowly but eventually snowballed into an avalanche of work that kept Dr. Gupta busy for years. By the time he retired from teaching in 2015, he had held roles of editor, assistant editor, associate editor, and member of the editorial board for thirty-three different professional journals. (See appendix A for a complete list.) Most of these roles spanned several years, and there was never a time where he wasn't supporting multiple journals simultaneously. One wonders how he stretched the short 24-hour day to accommodate this amount of reading and writing. Combining these roles with full-time teaching, being the father of two—soon to be three—young daughters, and traveling globally is a feat that few other professors undertake.

It is hard to imagine squeezing in any more activities in an already full schedule, but Dr. Gupta could and did, adding conference participation, becoming a visiting faculty member, and teaching during the summer. With an interest in a statistics conference that was going to take place in Canada, Dr. Gupta reconnected with Professor Tracy at the University of Windsor in Ontario, Canada. The two men had met each other at Purdue and had formed a good connection. Now Professor Tracy sponsored Dr. Gupta for research at the University of Windsor, which also enabled his participation in the Canadian conference. This established a very useful model for future trips during Dr. Gupta's long career, combining presentations at conferences with serving as a visiting faculty member. This small incident also highlights once again what a great asset Dr. Gupta's personable character has been for his international career. Even people he met as a student at Purdue became important figures in his later world travels.

The stay in Ontario was not Dr. Gupta's first visit to Canada. While living in Ann Arbor, he discovered that it was easy to cross the border to visit and to shop at the Indian grocery store in Windsor, Ontario. Alas, from time to time, the US customs at the border would question some of the perishable items he bought. However, rather than sacrifice them, he cleverly would turn around, eat the items, and then come back to cross the border with the contraband goods stored safely in his stomach.

Summer is always a welcome time for those in the teaching profession. While some new professors might look forward to the summer as a chance to relax, that was never a goal for Dr. Gupta, whose commitment to work was unceasing. Instead, he took on a summer position at Wright-Patterson Air Force Base in nearby Dayton, Ohio, continuing a relation with the Air Force that had started in Arizona. At Wright-Patterson, he worked with Dr. P. R. Krishnaya, conducting research and offering courses to the cadets.

Publish or Perish

Leaving the University of Arizona and coming to the highly driven atmosphere at the University of Michigan was metaphorically like hopping onto the back of the mythical swift-flying bird Garuda. The speed at which Dr. Gupta turned out research paper after paper while at Michigan was a major shift from his prior publishing efforts. He recalled, "I was at Michigan from 1971 to 1976. The highlight of those Michigan years was the pressure I got to publish some research papers. I usually wrote them with other people." In the five years that Dr. Gupta was at Michigan, he published singly or in collaboration more than thirteen papers. No doubt, the early exercise of turning out paper after paper during that period created a rhythm that he sustained for a long time after. The volume of Dr. Gupta's subsequent published research is somewhat unprecedented among his colleagues in statistics and one of his key contributions to the field of statistics.

The University backed up their interest in research by providing an environment that enabled the exchange of ideas and projects. Dr. Gupta recalled, "Michigan had a lot of money, so every week, they used to have somebody coming to the department. We ended up talking, and we'd write a paper." Often the collaboration would establish a mutually beneficial relationship. One paper, for example, was in collaboration with Vijay Rohatgi, "Estimating the Restricted Mean of a Continuous Population," predating their later work as colleagues at BGSU.

Three main collaborators during Dr. Gupta's time at Michigan each illustrate a unique aspect of his working style. The first three papers in 1972 were in collaboration with Dr. Zakkula Govindarajalu. "Raju," as he was known, was originally from Chennai, India. He had started the Department of Statistics a few years earlier at the University of Kentucky and met Dr. Gupta during a trip to Michigan to offer a series of talks. Dr. Gupta shared, "We got to talking about things and wound up writing papers together. We had a lot in common. In addition, his wife, Mrs. Gayatri Govindarajalu, happened to be a Gupta." Gayatri, much like Meera Gupta, was a well-educated and professionally active woman from North India. The Govindarajalus became good family friends of Arjun and Meera, and they exchanged visits to each other's homes over the years. Describing his work with Dr. Govindarajalu, Dr. Gupta explained,

The first two papers have to do with classification rules. Nobody had done that before. He was a nonparametric statistician, and I was a parametric statistician, so we combined the two and applied it to the classification. It was well received.

The collaboration with Dr. Govindarajalu epitomizes the way that Dr. Gupta successfully brought his heart into his work. Their warm personal relationship enhanced their ability to work together on abstract technical papers. Their collaboration also points to the openness and flexibility of mind that Dr. Gupta brought to his research. For some parametric statisticians, it might be a breach of "religion" to work with a nonparametric statistician. Yet Dr. Gupta was never sectarian about anything he encountered, and particularly not in statistics. Rather, the differences in their fields brought out his keen sense of curiosity to discover a way to blend two different approaches.

While Dr. Govindarajalu and Dr. Gupta had much in common, another collaborator, a Korean-born professor named Dr. Bock Ki Kim, could not be more culturally different. But as Dr. Gupta shared, "he was a statistician, and I was a statistician, and we joined hands together." This collaboration exemplifies Dr. Gupta's general stance that the universal language of statistics transcends cultural differences, a stance that found even greater expression in his later international career. Dr. Gupta and Dr. Kim published at least one more paper together after they both left Michigan in 1976. Dr. B. K. Kim went on to teach statistics at Memorial University of Newfoundland, Canada until he passed away in 1984.

The third collaborator during the years at Michigan was Dr. A. P. Basu, a faculty member at the University of Missouri. Dr. Gupta explained:

He came to visit me and we started talking about what I was doing and what he was doing, and we ended up collaborating. I visited him and he visited me. We worked together on papers during personal visits.

This particular collaboration points out the circumstances under which Dr. Gupta did his early research. In the mid-1970s, computer technology was still quite new, email was in its infancy, and online collaboration did not yet exist. As a result, their collaboration happened in person. Everything was handwritten and then typed. Under these conditions, the volume of Dr. Gupta's research while at the University of Michigan is even more impressive.

Expanding the Circle of Family and Friends

"You can see your wife and new baby now," the nurse announced, as she approached Arjun Gupta in the hospital waiting area. As Arjun was led in to see his third daughter, Nisha, nestled in Meera's arms, it probably did not yet sink in that from now on, he would be surrounded by four very strong women, with no male voice in the family as a counterbalance. If life in India had been patriarchal, that was now off the table for the Guptas! It was April of 1976, and by this time, Meera Gupta had decided to put aside her career aspirations in public administration to be a full-time mother taking care of Alka and Mita. With the addition of Nisha, Meera had her hands full running the household and raising three active daughters. While Dr. Gupta worked long hours to keep up with the University of Michigan pace, Meera ensured everyone, including her husband, had a supportive and adaptive environment, with lots of good homemade food. In addition, Meera led the way in creating a warm and welcoming social environment in their residence, hosting the many new friends Arjun and Meera would inevitably invite home.

The Guptas forged many long-term family friendships during their stay in Michigan. As Alka shared, "It feels like a lot of social fabric was there." The University of Michigan had more Indian students than the University of Arizona, partly due to location and partly because the College of Engineering was well-known for its excellence, attracting international students. A few people at Michigan came from Arjun's home state in India, Uttar Pradesh, such as Subhash C. Goel, who became a close family friend. Subhash, who was in the Department of Civil Engineering, received his undergraduate and master's degrees at Roorkee, the same city where Arjun's brother Vinod lived and only an hour's drive from Purkazi. Another family friend, Suresh Sanghal, worked in a private company in Ann Arbor. Arjun and Meera also met and stayed in touch with the Kumar family, who was from Meera's hometown, Agra. Gopi and Usha Jindal also became like extended family members during the Gupta's time in Michigan.

"Ann Arbor was a big deal, no doubt about it. And I enjoyed it," said Dr. Gupta. But in 1976, it was time to make a change. An opportunity to join the nascent statistics program at Bowling Green State University was hard to pass up. Leaving good friends behind, Arjun Gupta and his family prepared for one last major move, one that was only an hour's drive from Ann Arbor.

~ CHAPTER SIX ~

PROFESSOR GUPTA

In the winter of 1976, Raju (Dr. Govindarajalu) visited Michigan on the invitation of Dr. Gupta to present a lecture to the students and continue their research collaboration. Raju's next stop after Michigan was Bowling Green State University (BGSU). "Why don't you join me there?" Raju said during one of their working sessions together. That was all Dr. Gupta needed to hear, and he quickly said yes. He had already heard much about the Department of Mathematics at BGSU, which had managed to lure the famed Eugene Lukacs out of retirement a few years earlier to work there. A visit with Raju would be an easy way to check out the school.

As it turned out, this innocuous and spontaneous decision proved to be extraordinarily fortuitous. Arriving at BGSU, Dr. Gupta was introduced to the head of the Department of Mathematics, Dr. Terwilliger.

"Perhaps you've heard," Dr. Terwilliger began, "that Lukacs is retiring again, and we are looking for another statistics professor."

Dr. Gupta indicated he would certainly be interested in applying and promised to continue the conversation after he returned to Ann Arbor.

By now, Dr. Gupta's reputation was sufficiently impressive, due in large part to his many research papers and his unique contributions to multivariate statistics. BGSU offered him a promotion to associate professor with tenure, on a track to full professorship. Moving there would also provide the opportunity to serve as an architect of a new Statistics Department at a large university.

The faculty at BGSU was very supportive of Dr. Gupta joining the department. When Dr. Gupta interviewed for the position, the mathematician Dr. Motupalli, who was on the hiring committee, shared to a mutual friend Dr. Vasudeva, "Nobody in the department could match the credentials that Gupta is bringing to this department."

One of his former PhD students Dr. John Carson related that they had to convince Dr. Gupta to come to Bowling Green, adding, "After all, why would anyone want to come to Northwest Ohio?" Dr. Gupta recalled that "John Carson wanted me to come so that he could do his PhD with me. He was already a statistician at the time at an oil company called Marathon. He was my second or third student."

When Dr. Gupta accepted the offer from BGSU, his youngest daughter, Nisha, was still only a few months old; Mita was almost five; and Alka was six. Dr. Gupta finished teaching the semester at Michigan, packed everything up, and the family arrived in Bowling Green, Ohio, during the summer break of 1976.

Prior to their actual move, the Guptas made several trips from Ann Arbor to Bowling Green to search for a suitable place to live. However, this time, rather than searching for a rental house as they had in previous locations, they began looking in earnest to buy a house. With three young children to raise and a job that offered tenure, Dr. Gupta felt it was time to plant some roots for his family. A new acquaintance, Dr. Vasudeva from the English department at BGSU, described how he helped with the house search.

Since they wanted to buy a house, I took them around the town, and we looked at some neighborhoods where some houses were on sale. In a way, I was their guide in getting to know the town. Once they bought the house, then we had mutual invitations for dinner.

The Guptas ultimately located and purchased a newly constructed home with a spacious yard where Arjun Gupta carved out space for quite a significant vegetable garden, which he dutifully watered everyday upon returning from the office. Mrs. Gupta always thought that his diligent care of the garden was linked to his memories of the mango farms of his childhood in Purkazi. The family remained in the home for the next twenty-six years, then Dr. and Mrs. Gupta commissioned another house to be built—one with an even larger space for entertaining.

Department of Mathematics and Statistics at BGSU

The culture at BGSU has been described as easy-going by several former PhD students, who noted that despite the relaxed atmosphere, people still worked hard. The campus is compact, and one can easily walk from place to place, often coming across friends and colleagues when going from one building to another.

When Dr. Gupta arrived at BGSU, the Department of Mathematics, headed by Dr. Terwilliger, was in the process of change. In 1972, Dr. Terwilliger had invited Dr. Eugene Lukacs to come out of retirement and join BGSU as part of the initiative to establish a PhD program for statistics. Dr. Lukacs was a Hungarian-born statistician, educated in Vienna, who emigrated to the US in 1938 just after Germany annexed Austria. He became the director of an important statistical lab at Catholic University in Washington DC, which hosted prominent statisticians from around the world, including Sir R. A. Fisher, the father of modern statistics.

Upon accepting the invitation to come to BGSU, Dr. Lukacs also brought with him two other statisticians, Radha Laha and Vijay Rohatgi, whose books were used for statistics courses. For the next four years, Dr. Lukacs was instrumental in designing a graduate statistics program. His return to retirement in 1976 created a void that Dr. Gupta ably filled upon his arrival at the University.

Dr. Gupta participated in the advisory committee that was tasked with deciding how to move forward with a statistics department. As with many other notable schools, including Harvard, University of Chicago, and others, the discussion revolved around whether statistics should be a separate department of its own, or remain integrated with the Department of Mathematics. Ultimately, it remained part of the Department of Mathematics but equal in status. It then took a little while to get everyone on board with changing the department's name. Dr. Gupta commented, "It was quite a struggle. Somebody said we could call it the Department of Mathematical Statistics, but there's no such thing as mathematical statistics. It is statistics." However, a convincing factor for including statistics in the department name was that the federal government, academia, and industry were now offering jobs in statistics. Dr. Gupta pointed out, "If the students didn't have a degree in statistics, it would be hard for them to get a job." Eventually the Department of Mathematics became the Department of Mathematics and Statistics, with a separate statistics program committee, which Dr. Gupta headed for the next three years.

After two years as an associate professor at BGSU, Dr. Gupta was awarded full professor status. It was quite competitive, he said. "I worked hard and based on my record, it was recommended and I received it." Although publishing a book was not a condition of the appointment, it certainly added to his qualifications that by this time, he had already published his first book, *Parametric Distributions.*

Becoming a full professor freed up some energy and enabled his expansion into travel in addition to the continued teaching, research, and collaboration. This filled Dr. Gupta's schedule for the next eight years or so. Then in 1985, he became the department chair, only the second statistician to do so up until that time. His appointment was a recognition of the importance of statistics in the Department of Mathematics and Statistics, since the mathematicians had historically occupied that role. Dr. Gupta taught one less class to free his time for the administrative responsibilities that came with being the chair. Although he enjoyed the role, two years was sufficient. According to Mrs. Gupta, when they asked him to continue, he turned it down, saying something to the effect of, "I can advise you, but I don't want this."

Leaving behind that administrative role cleared the way for Dr. Gupta to once again dedicate his undivided attention to developing the statistics program both at BGSU and globally through his teaching and research, which he did for the remainder of his long career at BGSU.

Although he left the role of department chair behind, Dr. Gupta was a natural leader, regardless of the official title that he held. He was sought out by students as well as colleagues for advice, which they took to heart. If there were any problems, people in the department would say "Let's go ask Dr. Gupta." At Dr. Gupta's retirement party in 2015, his colleague Dr. Albert spoke of this.

> In terms of leadership of our department, of course, he was chair for a while, but also, he was the main leader of the stat group of the department. We've had our ups and downs, but I think he's always offered strong leadership. I think we have a good model to follow in terms of Arjun.

A similar observation was offered by Dr. Gupta's colleague Dr. John Chen, describing

> ... what we call Dr. Gupta's spirit. You know, every time the statistics group would have an issue or have a problem, he would always give some very insightful opinion or guidance, and give directions, like how to get funding for our students or how to get students working in different topics. I think he was amazing. I was lucky enough to join the department and learn from him.

Conference Organizer

As a leader in the department, Dr. Gupta also actively engaged in organizing statistics conferences at BGSU. Conferences have a dual effect of bringing talent to the school while making the school more visible to others. One conference of particular note was the first Eugene Lukacs Symposium that Dr. Gupta helped initiate in 1990. This symposium, in conjunction with the Lukacs Distinguished Visiting Professor, annually draws renowned statisticians to BGSU, such as C. R. Rao, who held the Lukacs visiting professor position in 1997/98. The conferences impact both the people who attend as well as the people create them. At Dr. Gupta's retirement party, John Carson jokingly shared,

> Well, if you hadn't created the Lukacs symposium, I wouldn't have met my wife, Paloma. She came in 1999 with G. P. Patil to work on composite sampling. Believe it or not, that was the biggest way you helped me. The PhD was a second.

His words drew appreciative laughter from everyone, including his wife, Paloma.

Organizing a conference involves a great deal of advance preparation, sending out the call for participation, creating the agenda, and attending to logistics. Dr. Gupta took all of this on with his usual zeal, commenting, "It was part of the life of the professor, and I enjoyed it. As the organizer, I got to decide who comes to the conference." Of most significance is that Dr. Gupta was a visionary regarding conferences and the possibilities they offered. Dr. Jim Albert shared this anecdote:

> When I arrived in '79, Arjun had this plan for having a conference at Bowling Green on the bootstrap, and he was inviting Brad Efron to speak. Efron is known for proposing the bootstrap resampling technique, which had a major impact in the field of statistics. This seemed crazy, like, how would you possibly get Brad Efron to come to Bowling Green to talk about bootstraps? Well, Arjun got research money, and it all worked out. And Brad Efron brought along a young assistant. His name was Persi Diaconis, who is now a very famous person at Stanford University. Persi was the after-dinner speaker. So, I think what Arjun showed me is that you can dream big, really grand, big things.

"Dream big" is an apt summary of the way Arjun Gupta expanded the world of statistics both at BGSU and around the world. There is a wonderful saying that the best way to make your dreams come true is to wake up. Arjun Gupta didn't just dream; he would wake others up to possibilities. He himself made things happen: conferences, international exchanges, publications, lecture series, and more, all part of his legacy to others.

Dedicated Teacher, Successful Students

Arjun Gupta's decision to pursue a teaching career upon graduating from Purdue clearly reflected the lasting influence of his father, who held the view that education was of primary importance. Arjun had completely internalized this message by the time he earned his PhD, and his commitment to education, both as his profession and its value for others, gained its full expression at BGSU.

In the newly created graduate statistics program, Dr. Gupta began working with selected PhD candidates, ultimately becoming the advisor for a multitude of students, thirty-three of whom achieved PhD status. (See appendix B.) Dr. Albert shared, "This is remarkable. I think he's always worked really well with students. We really have big shoes to fill in terms of mentoring students. I think Arjun did a really good job of doing that."

Arjun Gupta was always teaching, whether in or out of the classroom. He carried himself as a professor and dressed for the occasion, even when sightseeing or relaxing. Up until only a few years after his retirement, he wore a suit and tie every day. His friend Dr. Vasudeva observed,

> He is, in my judgment, pretty well-dressed. He impresses us as a person who looks like his distinguished career in academics. He dresses like an academic in a good but not very showy way. We should mention that he's very impressive, even physically. He attracts a person or impresses a person as well-dressed, tall, and handsome.

But it was not the tie nor his good looks that made Arjun Gupta a successful teacher. He has always pointed out to his daughters, "Knowledge is power." Ultimately, his teaching success comes down to this: "It doesn't matter how I'm standing or what I'm wearing. In the end, don't forget, the main thing is I knew more than the students." His knowledge of statistics was unsurpassed, as he continued to read, learn, and evolve as the field itself evolved. In addition to class preparation, his daily schedule always included reading journals and doing research. As he shared, "Teaching is an ongoing learning process." One of the memorable jokes that Dr. Gupta told his children is, "What do you call the student who stays on campus forever? A professor."

His competence in the field of statistics is complemented by, as noted often by his colleagues, personal characteristics that make him an extraordinary teacher. He loves to tell jokes, which always opens the way for great communication. He is an excellent listener, he is sincerely interested in others, and he is approachable. His niece Jyoti described her uncle's quality of listening and being interested in others.

> He is one of the few people who would be interested in what you are doing, especially academically. He was very, very into academics. He still is. And I always remember him saying, "OK, so if you're going to do law, what are you going to do? What is your purpose?" He was really helping you hone down on that purpose. This was one thing I think was really valuable.

His nephew Pankaj had a similar observation.

> When he met my kids, I prepared them, saying he was going to ask them questions, because he cares and he's interested. When he met my kids, it was always very targeted, like an interview: "What courses are you taking? What grade are you in? What do you like? What are you doing? And how are things going?"

Dr. Gupta's interest in others was a great asset as a mentor for his PhD students, and his friendly personality made him easily accessible to his students. One of his former students, Dr. Jen Tang, offered this:

> He's always friendly. You know, you can knock on his door, and he's always there, even on Saturday. I teased him at his retirement party. I said, "Now that you retired, I'm going to show up on Friday morning so that you don't have to come on Saturday."

The term *open-door* can sound trite, over-used, or hackneyed, except that in Dr. Gupta's case, it was literally true. At school, his door was open to students every day, who made best use of his generosity. Mrs. Gupta vividly described the scene at his office.

> I used to go to his office and there would be all kinds of numbers on his blackboard. I said, "Why don't you erase this?" He would respond, "Oh, no, don't erase it. We are still working on it. Don't erase this; the students will come in next week, and we'll work on solving it."

Dr. Gupta's good friend Dr. Bansal, a physician who lives in Lima, Ohio, and worked for a time at the BGSU health department, commented on his friend's character:

> He is a very dedicated man to his profession, and a very serious teacher. He would never shortchange the profession. He gave the best to it that he could, which is typical coming from India with his family background, like mine, where your parents instill the value that you just do your best at it. This is what I saw him doing. With his students, he was very flexible. The students said he was a very approachable

person. He never, ever said a negative thing about his profession or a negative thing about his students or about the college. He is always a very positive man.

It would be impossible to teach any branch of mathematics, including statistics, without a real, heartfelt connection to mathematics and the pure language that it represents. Dr. Gupta knew this quite well.

> You have to be good at mathematics and enjoy it. And enjoy the mental challenge of it, too. You can learn it, then you have to practice it, and you have to live your life with it.

And this is precisely what he did. Looking back on his teaching career, he commented,

> I prepared all my lessons; I spent hours preparing the lessons ahead of time. From daybreak to sundown, I was working. My work was all encompassing.

His family shared that he was in the office Saturday and sometimes Sunday as well, while also making time for the family. Even after he retired, he continued to help his PhD students, with some of them even stopping by his home to seek advice.

Dr. Gupta's skill as a teacher came from dedicated effort, his willingness to learn, and his keen power of observation. He said,

> I'm not sure if I'm a natural-born teacher, but hard work and will has supported me. I learned to teach with great difficulty. My own professor, K. C. S. Pillai, was my role model. I also had excellent professors in India at Poona and Banaras, and very good teachers at Purdue.

When asked about his teaching experience in India compared to the US, he made an interesting observation:

> Although that was an Indian teaching style, not Western, fortunately, in mathematics, it doesn't change much. You have to convey the material. And most of it is through the blackboard. Mathematics is the same in any language.

The appeal of Dr. Gupta's teaching style certainly crossed cultural boundaries. Dr. W. J. Conradie from Stellenbosch University in South Africa relayed his own experience during his sabbatical at Bowling Green in 1986:

> In those days, Professor Gupta was the typical mathematical statistical "chalk and talk" teacher. I enjoyed it. I found it very stimulating to sit in his classes. I experienced him as a very thorough teacher, who was really going out of his way to explain difficult mathematical statistical theory to his students by using a piece of chalk on the board.
>
> If I remember correctly, there was a very nice, open atmosphere in the class. He knew all the students, and he talked to the students in the beginning. He asked questions, and they asked questions. There was never a sort of wall between him as a teacher and the students. I learned from him how to formulate an advanced theoretical statistical research problem and then how to apply your mind to tackle the problem and to start to solve it.

Dr. Gupta also became a role model for those PhD students who went on to become professors themselves. Speaking about Dr. Gupta's teaching style, Dr. Tang shared,

I really liked his teaching. In the good old days, you would come in, he would lay out the problems, and then write on the board. Then he'd say, "Any questions? That's not right? Ok, let's erase it and start over again." I think that is a good method. You are learning the process. You are not sitting there just to receive information. You are learning the process and you are learning critical thinking, the logic. I really like his way of teaching. I still keep his lecture notes in multivariate analysis.

Dr. Tang shared that he still uses what he learned from Dr. Gupta.

One example I really remember from him: You have several variables, random variables, and each has a normal. You put them together and can get a multivariable normal, but it's not necessary. But on the other hand, if you have a multivariable normal, then each individual variable has to be normal. I always use that example in my own teaching.

Dr. Gupta did not limit his help to presenting classroom material. He has an innate sense of what might be of benefit to others in general. Dr. Conradie elaborated on this:

Due to his leadership in his department at Bowling Green and his status at that point in time within the international statistical community, it was possible for me to meet and mingle with some of the leading researchers in statistics and the broader academic world. He always came to me and said, "Well, there's a seminar at the University of Michigan. Don't you think you should go there?" And so on. He assisted me in building my own network. At that point in time, I was only thirty-four years old, at the beginning of my career. My interaction with him through the years, especially when I was a young academic, contributed enormously to my future career as an academic. It exposed me to the international statistical community, and it forms a significant part of my journey through life.

Dr. Gupta also contributed to his students' success by offering them the opportunity to collaborate with him on publishing research, modeling his own experience with Professor Pillai. Collaborating with one's professor is a compelling way to instill confidence and also helps the student establish himself or herself in the field. Dr. Gupta spoke admiringly about some of his PhD students who collaborated with him.

> Tamás Varga was outstanding. He was smart and he worked with me. Then he went to Hungary and became an insurance professional. He collected material and persuaded me to work with him, and we wrote a book, *An Introduction to Actuarial Mathematics*.

> Jen Tang was from Taiwan and came to the United States as a student to get his PhD. Jen Tang was brilliant. You give him a problem, and he would come back with a solution in two or three days, ready to discuss it and do some work.

Dr. Tang went on to become a professor at Purdue, crediting his appointment there to his work with Dr. Gupta.

> During a ten-year period, we published eight or nine papers. That was really a very critical point, as I think those eight or nine papers helped me to put my foot into Purdue. All my papers at that time were with Professor Gupta. I had a couple on my own but most were with him. I really

appreciated his help with them. That really helped my career.

Dr. Tang was involved in Dr. Gupta's later trip to Taiwan and still stays in close contact with his former professor.

Another student and collaborator was Jie Chen. Dr. Gupta shared,

> She was from Taiwan and a brilliant student. She came here and wrote her thesis. She was good at mathematics, and we wrote some papers. Then we wrote a book together, *Parametric Statistical Change Point Analysis with Applications to Genetics, Medicine, and Finance.*

An old adage states that a sure sign of a great teacher is that his students surpass him. There is no finer testimony to Dr. Gupta's skill as a teacher than the success of his students. When asked to reflect on his teaching career, Dr. Gupta shared, "Some of my students have done better than I have done, and that's a big satisfaction. They worked really hard."

For Dr. Gupta's retirement gathering in 2015, many former students sent comments on his contribution to their success. Here is a small sample of what his students said:

Tamás Varga (1990), wrote:
> When I look back at my years at BGSU, it is clear that you were the most important person for my academic development all that time. In my first semester I took your course "Multivariate Statistics." That course made me so interested in multivariate analysis that I decided I wanted to write my PhD thesis in that field. Fortunately, you accepted me as your PhD student, which started our common work

and research. You guided my preparation for the thesis very professionally. I recently heard on TV what makes a good reporter: "Ask the right questions and demand the right answers." I think this can also be said about the way you acted as my thesis advisor. You always encouraged me to write papers and books. Your encouragement did not stop even after I received my PhD and left BGSU. This resulted in a number of papers and books I can be proud of. Thank you for the constant help, support, and encouragement I received from you and for the friendly atmosphere which always surrounded you.

Grzegorz A Rempala (1996) wrote:
It has been many years since we spoke last but that does not mean that I do not think about you and the Bowling Green years often. Particularly now, when my own children look at graduate schools, I often return to BG memories and recall the early research projects which I pursued with your and Gabor Szelely's ... help. You have always been a great inspiration for me and I will be forever grateful for the opportunity to be one of your students. Thank you with all my heart. I know that you will likely continue to work and contribute to statistical literature, because one never retires from one's true passion.

K. P. Asoka Ramanayake (1999) wrote:
I'd like to thank you for being my advisor, helping me and guiding me throughout my study abroad. I feel so privileged to be a student of such a famous statistician. After graduating from Bowling Green, I worked for ten years at the University of Wisconsin and then I joined the census bureau and worked there for a couple of years. And now I

am finally at the University of Colombo at the statistics department.

Mentoring Others

Dr. Gupta was not only a support for his direct PhD students, but his particular expertise in the field of multivariate statistics led others to his doorstep to learn from him. He was invited to be the examiner of many theses from Canada, Brazil, India, Iraq, South Africa, United States of America, and Taiwan. Dr. Conradie from South Africa recalled his own experience.

> Professor Gupta was the external examiner of my dissertation. The title of the thesis was "Aspects of Multivariate Complex Quadratic Forms." Professor Gupta was one of a small group of people that did research in that field of multivariate complex distribution theory at that stage. I started my PhD at University of Cape Town, where my supervisor was Professor Cas Troskie, who knew Professor Gupta for many years. He sent my thesis to Professor Gupta as an external examiner.

Similarly, Dr. Gupta was approached by a university in Switzerland to be an external examiner. His friend Dr. Bansal commented on this.

> Arjun never bragged about his contribution. Never. Once there was a student in Switzerland who was doing a PhD on a topic that Arjun is the forerunner of, and the university invited him to give the exam to the student. When he came back from Switzerland, we discussed it. I said, "The Swiss were tremendously courteous, and they honored your subject by inviting you; apparently you are the top dog in

the subject. And the subject must not be that easy, otherwise anybody would be doing work in that." But he never had an air about himself. He never talked about it until I asked him about it.

In Retrospect

Dr. Gupta responded to a question about students today compared to the past, saying, "Students are certainly very different."

> I would like today's students to be more responsible in terms of what they are being taught, rather than just get a degree or diploma. There is a lot more distraction for students, too.

For students in statistics now who are researching, he added, "There is only one advice: Work hard, work hard."

When reflecting on his career, Dr. Gupta remarked, "If I were going to redo my life, I perhaps would be involved in much broader subjects, but still in relationship to educating people." In 1999, Dr. Gupta was honored as Distinguished University Professor by BGSU, a recognition that he richly deserved.

Roots in India, Life in America

Dr. Gupta now had tenure at BGSU, owned a house, and had three growing daughters being educated in the local schools. One might think that he was permanently committed to life in the US. Yet the motherland India was never far from his heart and mind even after settling in Bowling Green.

In the very early years, Dr. Gupta still had one foot in India, because his birth family was there. But it was more than that. His nephew Pankaj related,

He loved India deeply, even though that sounds like a cliché. Once we were driving on a long trip to drop something off in Cleveland to Alka. We had the discussion of where I would live. I was thinking of California where I had a few jobs potentially lined up. Chachaji said, "You know, Pankaj, at the end of the day, there are only two places in the world you want to live in, Delhi or the US." Even though he was very well-traveled, having gone everywhere, it was like there was no other place.

Many international students who come to the US to study grapple with the decision to stay or return after their studies. Dr. Gupta's good friend Dr. Jain related his own experience. "Originally, when I came to the US, I knew for sure that my plan was just to get educated and go back. I guess that was the case with him too. For most people, the idea was that you came for education, you got educated, and then you go back." He explained how that intention slowly changes:

> It's not a very easy decision, but I think the longer you stay, the clearer it becomes. Once you leave India for fifteen or twenty years, then it becomes kind of a different world. You can become a misfit there. But early on, you could go back. I could have just graduated and gone back, and there would have been no problem.

Even after the decision to stay is made, there is still a pull to India, Dr. Jain shared.

> You always think about whether it would have been better to go back. You always think that had you gone back, you would spend time with the family. So that's always there in the back of your mind.

In the 1980s, after being in the US for more than twenty years, Dr. Gupta's commitment to stay was put to the test when he received an unsolicited job offer from the University of Delhi. The University had reached out to Dr. Gupta, but as he related, "It was a two-way process. I wanted to go back to India, and Delhi was a place I could have done some work." At least three of his siblings and their families were living in Delhi, and it was very tempting to consider returning to India. Both he and Mrs. Gupta had relatives who had lived in the US for a few years and then returned to India to settle.

The whole family was involved in the discussion about whether he should accept the job. Mrs. Gupta described the process.

> We really thought of going. But then when we seriously thought about it, we knew that once you get used to the comfort of life here, it is very difficult. Especially when you have children here. Had we gone when there was just the two of us, it would have been different. But at that time, with three kids, how would we manage?

Alka was in the fifth grade at the time, and she and Mita had practical concerns: "Oh no, we won't have our Schwinn bikes there! Not to mention leaving our friends and our entire way of life!" In the end, Dr. Gupta turned down the offer. He said,

> The benefit of staying in the US was the freedom to work, the journals that were available, and the ability to do research. I absolutely would not have had that same opportunity if I went back to India. Teaching would have been the main thing.

Once they decided to stay, some practical things changed. Alka explained the shift.

> Before that time, whenever we used to buy big items, there was always the conversation that the appliance should be dual voltage. After we decided not to move to India, we didn't have to worry about those hard-to-find dual-voltage appliances. Even our car…sometimes we thought about getting a station wagon for our car, but we knew that we couldn't operate a station wagon in India. After that, we didn't have to think about it. So, our life got settled.

"Welcome Home," reads the sign above the exit turnstiles at JFK Airport, designated only for US citizens, who breeze through without a wait. Immediately adjacent, hundreds of passengers stand in long lines under the sign "Terminal 1 Arrival Immigration," waiting for interviews as non-US citizens. Each time the Gupta family arrived in the US from an overseas trip, Dr. Gupta and Mrs. Gupta separated from their children, following the arrow for "Resident Aliens" while the children followed a separate arrow to the US citizens line. For more than thirty years, Arjun and Meera remained Indian citizens, reluctant to renounce their Indian citizenship. When he traveled to India, Dr. Gupta could take his family through the line marked "Indian National" rather than the off-putting "Non-Indian" immigration line. Giving up Indian citizenship, he shared, "was very hard." A friend observed, "It is like giving up your mother. You don't give up your mother for another mother. You just can't do that. India is the motherland."

His good friend Dr. Bansal had pragmatic observations about the citizenship decision. He said, "The community Arjun comes from has a very strong sense of their own dignity. Also, his brothers were high-level officials in India, and there was a sense of belonging. He didn't have to become a US citizen because it didn't really give him any advantage at that point. As a green-card holder, he was getting promotions and moving up the ladder."

Finally, on July 5, 2001, Dr. Gupta officially became a United States citizen. He now has an American passport and during their travels, the family stays together as they go through the immigration lines re-entering the US, although it is a very odd feeling to return to India and go through the immigration line for "Non-Indians." Despite the passport, Dr. Gupta is definitely not a "non-Indian."

Very surprisingly, once when Dr. Gupta was asked if he would have done anything differently, he responded, "I might have stayed in India and done better. My brothers and sisters did very well in India." However, that is obviously not the whole picture. When asked if he was happy now that he was in America as an American citizen, he responded, "To me, no doubt that's true. I've been here for almost sixty years."

Nurturing the Indian Heritage

Turning down the job in Delhi reinforced Dr. Gupta's decision to stay connected with India as best he could. Every two or three years, he and his family spent time in India, often during the summer holidays, staying with their relatives. While the trip to India was quite a costly undertaking for a family of five, Alka points out that it was a conscious decision to prioritize it even though other families might say, "Wait, why would you do that?" However, Dr. Gupta's dedication to his brothers and sisters was unwavering, and he commented, "I shared that with my children, too. It was important." The trips clearly had a lasting impact on his daughters as they each have significant memories of their visits. Mrs. Gupta shared, "They really understand Indian culture and the closeness of the family. Even now they talk about the old days when we used to go to Delhi, so this is very fulfilling for us. The children have memories of back home." The trips to India also created lasting bonds between both the older and younger generations. Some of the cousins have come to the US to visit and they all still keep in touch with each other, in part, due to the lasting relationships they built through those early family experiences and trips to India.

Delhi was the obvious location as a home base for the visits to India. Most of Dr. Gupta's siblings lived there, and Mrs. Gupta grew up in Delhi, where her ancestral home and family were. From Delhi, Dr. Gupta and his family would take trips elsewhere, such as to Purkazi, Bombay, or Roorkee, depending on where other siblings were living at the time. They also took trips to Agra where Mrs. Gupta had family. The big family gatherings were often at Bhaiya's (Arjun's elder brother) large house on State Entry Road, located in Connaught Place in the heart of New Delhi. Everyone converged at the house, which was located on an urban estate with lots of helpers and was in a central location near a big shopping area. Bhaiya's daughter Jyoti described the scene.

> It was so much fun. All the siblings and their kids would get together, so all the cousins got to hang out with each other. Everyone stayed in the house. We would be six kids in two rooms. While the older people would be busy talking, all the kids would get out. There was always food on the table, so we would just meet the parents for meals. Otherwise, we would be just walking around, watching movies, and going all over the place. We really looked forward to those holidays, which were usually in the summer.

Visits to India were, of course, a significant way for the Gupta family to remain connected to their Indian heritage. However, since the trips only occurred every two years or so, home life in Bowling Green also played a big role in transmitting the Indian culture to the next generation.

Immigrants from another country often face a dilemma: how to keep and transmit their home traditions to their children in a new land. Refugees fleeing from persecution, such as Eastern European immigrants to the US in the 1900s, often see assimilation as a form of self-protection. Yet many other immigrants prefer to maintain their culture of origin while simultaneously embracing the new culture. This was the choice that both Dr. and Mrs. Gupta made. Commenting on assimilation into the American culture, Dr. Gupta explained,

> It was a slow process—without knowing, we adopted the new culture and kept in touch with the old culture through letters and phone calls. Through me and my wife, the children stayed connected with the Indian culture, intentionally.

Dr. Gupta's friend Dr. Pradeep Gupta commented, "In spite of his being in America for fifty or sixty years now, Arjun is not Americanized. He has found a confluence of cultures. He has not left his previous culture. He still has a very strong imprint of that culture in him." Another friend, Dr. Bansal, had a slightly different observation about this: "We *(he and Dr. Gupta)* both are Westernized in a good, positive way, yes, but we also have our Indian roots that we hold dear. We don't allow a conflict between those two roots, because we find that Western culture has a very positive contribution to our lives." Family friend Dr. Vasudeva further observed,

> He has retained his faith despite the challenges that come with other faiths or with science. At one time, we talked about the contradiction between the role of the scientist and religious faith. The British citizen, Stephen Hawkins, when he died, said, "By the way, there's no God." Arjun

and I discussed that. Despite that conflict with science, I believe he has retained his faith. Recently we were talking about life, and he said he was born a Hindu and he would die a Hindu. That kind of thing defines him, though it may not come through in a social setting. But as a person of faith, he has retained his cultural and religious values despite having lived here for a long time.

In the small town of Bowling Green, the social life with other Indian families was a big factor in keeping the Indian culture alive. Family friends Dr. and Mrs. Vasudeva shared many common interests with the Guptas, getting together on some of the Indian festival days. Dr. Vasudeva described a typical social setting.

> There was not much difficulty in getting all the faculty of Indian origin together in one house. This small group in a small town gave us an opportunity to have a social life when there was not much other diversion like theater or cinema halls or opera.

The young children of the faculty were always included in the gatherings, forming their own lasting friendships.

Dr. Gupta did not compartmentalize his personal life and professional life the way others might, and he openly shared aspects of his life that were meaningful to him, such as his Indian heritage. His former student Tamás Varga wrote to Dr. Gupta at his retirement,

> During the years we developed a good personal relationship. You shared many things about your cultural heritage with me. So, while staying in the US, I also learned a little bit about various aspects of India like the Hindu religion, the Sanskrit language, and Indian mathematicians.

Dr. Vasudeva recounted,

> Arjun is very outgoing in terms of the Indian diaspora. He wants to know everybody and offer his support. As the Indian faculty group grew larger, we started getting students in the school of Indian origin, and we would have a meeting once or twice a year on the college campus where students could come. The faculty families would bring food so that students could enjoy homemade Indian cuisine and so on. It has been a very rewarding and fulfilling part of our life here.

"*Ka, kha, ga, gha*…Now you say it," their mother said. And five-year-old Mita and seven-year-old Alka would repeat it. "Now write it in your notebooks," Meera Gupta directed, and they did. When Nisha was old enough, she too joined them, although begrudgingly, according to her recollection. Seated around the kitchen table, Mrs. Gupta gave Hindi lessons to her children as soon as they were old enough to learn. Mrs. Gupta explained,

> They can speak fluent Hindi, although the real Hindi accent is not there. But I'm glad that they can talk and they can understand. I taught them when they were very little. Every night I would sit down with them with a Hindi alphabet book and I would teach them. And then I did some dictation and they would learn it.

The girls also learned the Hindu rituals simply through observation, as they watched their parents perform traditional practices such as a morning *aarti*, or worship, during the Diwali holiday. Some practices were more directly instilled; for example, when their father drove them to school, the girls practiced their recitation and memorization of the *Gayatri* mantra, one of the most sacred mantras in the Hindu tradition. Later, the Guptas were actively involved with other families in creating a Hindu temple in nearby Toledo, so that all the Indian families could gather and continue the traditional practices.

Arjun Gupta also shared his love for Bollywood movies, which were the bread-and-butter, the *naan-and-ghee*, of his early life in India. The family regularly watched Bollywood movies all through the '80s, which exposed the girls to the Hindi language while immersing them in Indian culture for an hour or two. However, both Arjun and Meera made sure that the daughters were equally comfortable in the American culture that they lived in. Nisha shared, "I appreciated that the Indian and American cultures and language were seamlessly blended in our home—Friday evening watching the latest Bollywood release on a VHS, and the next dragging my parents to the record store to buy the new Depeche Mode tape." She humorously added,

> In fact, one of my college application essays was entitled "You are what you eat," and I waxed on about how I "am" *dal* and *roti*, but also burgers and fries.

Dr. Gupta's nephew Vishwanath Prakash was quite impressed with the way the girls integrated India into their life. He said,

> His children are so remarkable. These three are Americans, born and brought up there. So why should they learn the Indian language? Why should they be able to converse in Hindi? But that was his influence. He's a unique person. He's very strong in his beliefs, not rigid, I'm sure, but has very strong feelings for the motherland, people, the family, and these ideals. That is how this family is what they are. It's not easy to survive in a foreign land, emotionally,

physically, yet it is such a beautiful bond to have these three daughters dote on their parents.

In short, through travel, food, movies, music, and language, Dr. Gupta's three American-born daughters received the gift of their parents' motherland right in Bowling Green, Ohio.

A Daily Rhythm

Arjun Gupta's professional life at BGSU combined teaching, mentoring students, researching, publishing, and traveling. His work life was all-encompassing, leaving little room for hobbies. Teaching is a particularly under-appreciated profession, as there are always classes to prepare, papers to grade, students to meet with, and so forth. As Dr. Gupta said, "I think it's very difficult for non-teachers to know how much time it takes." His daughter Mita shared,

> He was either working, or with us, or he was traveling. He would come home with his briefcase, and my mother would jokingly say, "Oh, he's with his mistress," because he was very dedicated to that.

However, his home life also received his loving attention. It was a bit of a balancing act but it all came together, thanks in large part to Mrs. Gupta's partnership.

Arjun Gupta had very full and long days, leaving the house daily at 7:30 am and returning home promptly at 5:30 pm. Each morning, after he woke, he offered a candle and a prayer at the altar next to his bed, then went downstairs for tea and a piece of toast. Usually, the *Today Show* was on with the morning news. The girls eventually joined him for breakfast. Alka shared her recollection of his daily schedule, which was exceedingly disciplined and punctual.

In the morning, he dropped us off on the way to school. He was home by five thirty. Then we ate dinner. We watched the news at six-thirty. Between dinner and news, if the weather was good, he would go out to the flower garden or vegetable plants to water or weed. During TV prime time, Dad would multi-task and review a journal article. Then we all went to do our homework. He took his briefcase out and worked on papers while we would be studying. Then we would go to bed, and he would come and visit us before we went to sleep.

On Saturdays, he would go to the office in the morning and come home by noon, carrying his briefcase filled with papers. He would bring home papers to work on each day because, he said, "that's the work I can do at home."

Despite the work-filled schedule, Dr. Gupta always included his family in the various activities that he enjoyed such watching movies, playing chess, and listening to Indian music, which continues to enrich everyone's lives. "As for chess," he explained,

> in the coffee room at school, there was a chess set, and I used it. Anybody would come in and I would play them. At home, I taught Alka first, and then Mita, who is the one who took it up.

Mita said, "We used to play and then he started to pay me money when I would beat him." When his grandson Arhaan came to Bowling Green one spring break, Dr. Gupta taught him chess, and now Arhaan and his grandfather enjoy a friendly chess competition as well.

One of the hallmarks of life at the Gupta home was and is the gracious, welcoming atmosphere for guests. Their home is intentionally large enough for entertaining guests. The Indian culture, perhaps unlike most other cultures, elevates hospitality from mere polite social activity to an act of worship. The Indian scriptures teach that the guest is God and should be treated with respect. This generosity is such second nature to those of Indian origin that they may not understand why non-Indians make a big deal about it. Yet when speaking with anyone—former students, faculty members, relatives, visiting faculty—about Arjun Gupta, his gracious hosting and open house always comes into the conversation.

Here is just one example: In 1986, when Dr. Gupta's daughters were in their teens, the whole family hosted Dr. Conradie, his wife, and his young children, ages six and four. Dr. Conradie said,

> I want to emphasize that my visit to Bowling Green was made possible by Professor Gupta. At that point in time, South Africa was in turmoil and especially white South Africans were not so welcome everywhere in the world. I appreciated that he just didn't ask any questions about politics and the situation in South Africa. He just accepted me and said, "You are more than welcome."

Dr. Conradie continued,

> My wife and I will never forget how kind he and Mrs. Gupta were to us when we arrived in Bowling Green with two small kids. It was the first time that we went overseas and to the United States. Everything was big and foreign to us. They were tremendously supportive to us and really went out of their way to let us feel at home. I can think of

numerous occasions that we, as a young family, were entertained at the Gupta home. We enjoyed delicious dinners there, and we will just remember that for the rest of our lives.

Colleague John Chen echoed the hospitality he received, saying,

> As an international scholar, I really appreciate his help, especially helping me to settle down in Bowling Green, finding the house, finding schools for the children, and finding all these little things. He and Mrs. Gupta both helped us with a lot of things. I actually didn't even know how to go to a conference. At the first conference that I attended, he actually took me there, and showed me how to do it.

Family friend Dr. Bansal observed that the Guptas were great at making meals at home that people from other cultures could enjoy, which was important given how international the network of visitors proved to be. Another friend, Dr. Vasudeva, commented,

> He very much had the feeling of being helpful to other people. I am impressed that he retains that kind of connection with what we could call immigrants. In earlier stages, when we don't know the ways of American life, it helps us when somebody is there to guide us. He has been a good guy to anybody who came.

Dr. Vasudeva shared a dramatic incident that went beyond just hospitality: A young couple of Indian origin had arrived sometime in 1979 at the University of Toledo, and he and Dr. Gupta had befriended them. The husband was confined to a wheelchair due to an earlier accident. One night, about 2:00 am, his wife called Dr. Gupta, saying that her husband had just passed away from choking. So, in the middle of the night, Dr. Gupta and Dr. Vasudeva drove twenty miles to Toledo to support the distraught widow, arranging for a doctor, funeral arrangements, and so on. Perhaps about ten hours later, the two men returned home. As Dr. Vasudeva related this incident, he said,

> This is an example of who Arjun is. His mind and heart: both are great. Very generous. Somebody could fault him for not showing off his scholarship, but that's an attractive quality in him, that it hasn't gone to his head. In everyday life, he remains a very good friend, a warm person, a social person.

The routine in Bowling Green defined and anchored Arjun Gupta's life for the next four decades, a life which included scholarship, research, and bringing statistics to the far-flung reaches of the globe.

- Photo Gallery -
Part II: America

Arjun Gupta as a student at Purdue, 1965

Newlyweds Arjun and Meera Gupta in New Delhi after their wedding, Dec. 1967

Part II: America

Meera Gupta with Arjun's sister Sarla (*Jiji*), and
Arjun's nieces and nephews, visiting the mango orchards near
Purkazi after the wedding, Dec. 1967

The young couple in Tucson, Arizona

Meera and Arjun Gupta
as tourists in Las Vegas, 1968

Dr. Gupta, Meera Gupta, Alka, and
baby Mita, 1973

Four brothers during one of
Dr. Gupta's visits to India

PART II: AMERICA

Four brothers on the lawn at Ram Nath's *(Bhaiya's)* house in Delhi in 1986.

Professor C. R. Rao in Dr. Gupta's office at BGSU; April 24–26, 1998, Lukacs Symposium, "Statistics for the 21st Century"

Dr. Gupta officially became a United States citizen, July 5, 2001.

Dr. Gupta used transparencies for teaching well into his later career.

Part III: The World - Trailblazer for Statistics

Even when another person seems a world apart from you, find a bridge.
Arjun Gupta

~ Chapter Seven ~

A Scholar with Heart

Undoubtedly, a defining feature of Dr. Gupta's career has been his dedication to statistics. Many colleagues and friends commented on this quality. His friend Dr. Vasudeva noted:

> He has a whole-hearted commitment to his profession that I would say, in my perspective, was the best characteristic I saw in him. His devotion and commitment to his field must be one of the factors that resulted in his achieving such distinction in his own field.

Dr. Gupta's contributions to the field of statistics through teaching, research, scholarship, and publications are intertwined with his travels and his participation in numerous statistical associations and conferences. Often after publishing a paper, he would be invited to present it at a conference, which would lead to meeting someone new with whom he might collaborate and publish. Or he might attend an association meeting and meet someone there who wished to collaborate with him. In this way, the wellspring of contacts, research, publication, and travel was always replenishing itself in an unending cycle of creativity.

Theoretical and Applied Statistics

It can sometimes be challenging to understand the world of statistics and to present a context for the field in very simple terms. A great deal of statistical work, including that of Dr. Gupta, can seem extremely theoretical and even esoteric. However, it has far more meaning than just an intellectual exercise. The origin of statistics is, of course, mathematics. Centuries ago, Galileo described the importance of mathematics: "The universe cannot be read until we have learned the language…in which it is written. It is written in mathematical language…" When presented with Galileo's statement, Dr. Gupta expressed his agreement.

> Without mathematics, the universe cannot be explained. It means that you have to have logic. Logic is needed for creation, re-creation, building, and such. The universe is not created randomly. I think God has a plan and he describes it through mathematics.

As a branch of mathematics, statistics is not just about numbers. Dr. Gupta patiently explained, "Statistics is everywhere. You can't found the universe, you can't found the country, you can't found the city, or even a department. You have to have statistics before you. Statistics goes beyond just addition and subtraction of numbers."

Unlike mathematics, however, the field of statistics had an uphill battle to find acceptance in the universities, especially in the US. In 1954, when Arjun Gupta was at Banaras Hindu University and just beginning to consider studying statistics, only nine schools in the US had statistics programs: Berkeley, Chicago, Columbia, Iowa State, Michigan, North Carolina Chapel Hill, North Carolina State, Princeton, and Stanford. Even the venerable institution of Harvard looked down on creating a Department of Statistics. It was only created in 1956 after one of their valued professors of mathematics and statistics, Frederick Mosteller, threatened to leave Harvard to go to University of Chicago to chair their statistics committee.

Nevertheless, despite its rough birth in academia, there is no doubt that statistics is now a necessary component in the world, answering complex questions posed by nature and humankind.

As statistics became established as a unique field of study, a separation arose between theoretical statistics and applied statistics. The two paths are quite divergent today, and they have been for quite a while. Dr. Gupta's friend Dr. Jain shared his experience when he arrived at Purdue to study statistics.

> I went there to major in statistics like Arjun, but after I got there, I found out that I was interested in applied statistics while Arjun was interested in mathematical, or theoretical, statistics. At that time, the Purdue Department of Statistics emphasized mathematical statistics. They had a master's degree in applied statistics but did not have a major for the PhD. So I ended up partnering with the School of Industrial Engineering. If I was a new student now, I could do a PhD in applied statistics.

Dr. Gupta's former student Dr. Jen Tang, who transitioned from theoretical statistics to applied statistics, explained the way the field of statistics looks today:

> There's a journal called *Annals of Statistics* today, which used to be called the *Annals of the Institute of Mathematical Statistics*. They publish only theoretical research results. You can still see a traditional, hard-core type of paper with no numerical examples. It's just all theorems. There is another journal called *JASA* (*Journal of the American Statistical Association*). They publish more about methodologies. There

are two sections: theory and methodology, and applications and case studies.

Dr. Gupta's focus has always been theoretical statistics, or pure mathematics. This purity shines forth in his published articles, which are elegant and beautiful even to someone who doesn't understand their content. Although he is thoroughly committed to theoretical statistics, Dr. Gupta has always maintained a strong interest in real-world applications for his work.

> It's always good to have some applications in mind. But with that, you have to formulate it, you have to describe it in mathematical form, and derive certain results mathematically. Then you can apply it to the field.

Thus Dr. Gupta's research has both theoretical and applied components. Some of his published papers and books may sound very abstract, presenting pure mathematical results on topics such as change point, skew, Gaussian and Non-Gaussian models, distribution theory, and so forth. However, many of his joint research papers bring statistics directly into the solving of real-world problems in areas such as agricultural science, stem cell research, drought, drug dosages, finance, insurance, biological research, stock market results, and others. As Dr. Gupta said, "I didn't have a preference about the kinds of papers I worked on. The only preference I had was that the collaborators should have a smart brain."

Dr. Gupta is keenly aware of how statistics directly affects people, such as using statistics to calculate the state boundaries. He wryly commented,

> How do you draw the state boundaries? With the politicians away, it's easy. But when the politicians enter the

field, it's more difficult. It's all based on statistics, but the politicians don't know statistics. They make up statistics. When I was at the United Nations, I did some consulting work, for example, in Thailand. But I was only a consultant, because the politicians have the final say.

When asked what he thought was the most significant application of statistics, Dr. Gupta discussed the census. "Every tenth year we have the census, which makes a difference for everybody. You count everybody—who is where and what they are doing." As if to underline his comment, shortly after he said this, the 2020 US census results were released, shifting the political landscape, which may ultimately have its own ripple effect in the US.

When Dr. Gupta reflected on what he might do differently if he were starting his career all over again, he offered some surprising and interesting observations about the field of statistics: "I would not pursue statistics. The subject is abstract and the outcome may or may not be that important." Dr. Gupta continued, "The applications part of statistics is hard to publish. The work that goes before that—the theory and the results that can be applied—is more important. The applications come later. First you have to have results."

In other words, there is not as much recognition for the applications of statistics, despite their important and direct impact on society. When asked how to remedy the system, Dr. Gupta replied,

> Well, it's hard to say, because when the PhD and other research is done, the theoretical part comes first. The National Science Foundation (NSF) does not support just applications. The money is allocated to NSF, and NSF then allocates money to different branches. It is more

important to find $E=mc^2$ than to look at applications. I could be biased, but I don't think that's okay.

Despite this interesting retrospective, Dr. Gupta's research, well-grounded in theory, has become the foundation of important applications in the world. His friend Dr. Jain explained,

> Arjun stayed academic, and he was very productive in writing a lot of papers and interacting with a lot of people. Those kinds of papers eventually became the seed for the applied mathematical use of statistics. It takes time for them to get to that stage but eventually it happens.

Dr. Gupta's specific area of expertise, multivariate statistics, forms the basis of important modern-day applications. According to one source, "*multivariate analysis* is what people called many machine learning techniques before it was so lucrative to call it *machine learning*." Machine learning, according to Dr. Tang, is basically a statistical estimation. "You train the machine to estimate something. Basically, the more data the machine has, the better it learns." Multivariate statistics is also used as part of data analysis in psychology. Dr. Tang shared, for example, that *Psychometrika* is the best applied multivariate journal using statistics.

In order to appreciate the impact and application of Dr. Gupta's research, it may be helpful to look more closely at the basis of his work before looking at the specific results he achieved.

Multivariate Statistics and Wilks' Lambda

Dr. Gupta's initial research in multivariate theory, which was encouraged and influenced by his PhD advisor, Professor K. C. S. Pillai, was related to a test statistic known as Wilks' Lambda. Discussing his PhD thesis, Dr. Gupta explained,

There was a fellow from Harvard named Schatzoff, who wrote his thesis on Wilks' Lambda. When I read it, I thought he had done only half of the work, so I worked on that. I came up with solutions that I published, and I made tables for it. The subject was the percentage points of the Wilks' Lambda. And that was the research for my degree.

The title of Dr. Gupta's PhD thesis was "Some Central and Noncentral Distribution Problems in Multivariate Analysis." While Dr. Gupta made many more contributions to the field of statistics, he would like to be remembered most for his work on Wilks' Lambda. As he stated, "I published the tables out of it, which are still there." But what is multivariate statistics, and what is the significance of Wilks' Lambda? Dr. Gupta's former student Dr. Tang offered a very detailed answer to these questions. For those readers who might be mathematically inclined, here is what Dr. Tang said in part:

> Wilks' Lambda is a key test statistic in multivariate analysis. Professor Gupta always told us that everything is multivariate. Statistics is all about the distribution of statistics, because if you know the distribution, you know the variability of your statistics. Then you can control the risk and the probability of making a wrong decision based on statistics. So everything is multivariate, and statistics is basically about distribution.
>
> In statistics, we are interested in the mean, the population average, and also the population variance. But in the multivariate setting, we're interested in tests about the mean vector—a vector of the means of several variables—or a covariance matrix of a vector of variables. And in the

multivariate setting, of all these tests, the most famous, most important test of statistics is called Wilks' Lambda.

Wilks' statistics is used in traditional statistics, but the difficult part was to find the distribution of that statistic in an easily computable form. Dr. Gupta, along with his advisor from Purdue, Dr. Pillai, were the first ones to derive the distribution of the test statistic in closed form. Not only the closed form but also the sum of a finite number of terms. At that time, back in 1963, they derived the exact distribution in a very closed form, with finite terms or some special cases.

This detailed background on multivariate statistics and Wilks' Lambda provides the context in which to view Dr. Gupta's outpouring of research and publications.

Research and Publications

Throughout his career, Dr. Gupta was unswervingly committed to research. As he once said, "Research accepts only one lover—you can't have two." At its core, research is an exploration of new horizons, which matches well with Dr. Gupta's sense of adventure. He explained, "In research, everything you do is new in our field. So the research is kind of surprising in that sense. You don't know where it's going to lead."

The process of mathematical research is extremely creative and differs, for example, from historical or archaeological research where one relies primarily on existing or physical materials. Mathematical research relies on the mind's ability to detect something that has not been postulated, and then one sets out to prove it. Mathematical research is also not passive. It is said that Einstein's breakthrough came when he was idle, sailing on a boat to clear his mind. Dr. Gupta's process is more like jumping off the boat into the water to see what he can find. It is about keeping one's mind actively engaged in the topic. He explained his method of research with great clarity:

> You are a member of two or three journals, and you read what was written, then you go from there. And you have to come up with something original, which comes from one's own understanding of what you are seeing: what you see, what can be done, and what they have done. It's not that easy, and it doesn't come every day. It comes off and on. But when you read something, the questions come automatically to your mind, and you just wonder if it can be done this way, or in some other way, or a better way. You are analyzing what other people have done. When you read it, that part of the brain picks up things that are not there.

Being original is part of the research process, but research is not done for its own sake as a purely personal, intellectual pursuit. It is not about research in isolation from others either. Sharing one's discoveries in statistics is an essential component. According to Dr. Gupta, "That [sharing] is the only thing we do as researchers besides teaching, which is the bread and butter. But research means exactly that—sharing, widely sharing."

The results of Dr. Gupta's research have been published in a variety of forms, including journal articles, book chapters, conference papers, and entire books. At first glance, the volume of his publication seems impossible to achieve: over five hundred publications between 1967 and 2020, which averages to almost ten papers a year for fifty-three years. Mrs. Gupta, observing his relationship to research first-hand, shared, "He kept publishing, publishing, publishing, and then it just consumed his thought process."

When he was asked how it was possible to produce so much, Dr. Gupta modestly pointed out, "Don't forget that I had thirty-three PhD students who did quite a lot of work." However, that statement belies the fact that mentoring thirty-three PhD students is in itself a remarkable achievement. Dr. Gupta's daughter Mita vividly described the process as she observed it: "It was a continual steady stream, like the Energizer Bunny. He just kept at it. Earlier on, it was a way to get established. Later on, it just kept going."

Dr. Tang offered another explanation for the volume of publications by Dr. Gupta: "It's an amazing amount of papers he published. It is not common. How that came about, for one thing, is that he knows a lot. And another thing is that he works hard."

The impressive list of over sixty high-caliber coauthors (see appendix C) is noteworthy for the diversity of cultures that are represented. This very global set of collaborators reflects Dr. Gupta's conviction that mathematics and statistics transcends the limitations of language and culture, and it is also indicative of the breadth and scope of his travels. How does one find a suitable collaborator? Dr. Gupta explained:

> First of all, anybody doing research has to establish themselves. Then students come and join you who want to do work with you, so the process is a continuing process. I did not seek out all the people who I worked with—they connected with me. I did not agree with all of them, so I said no to some. But those with whom I agreed, we did some work together.

In addition to publishing articles for journals and conferences, Dr. Gupta also has published over twenty books, many of which continue to be in use and relevant in the academic field. (See appendix D for a list of published books.) During an interview, his former student Dr. Tang held up a book, stating, "I still have his book that he gave me, *Multivariate Statistical Analysis*. I threw away a lot of books, because I'm close to retirement myself, but I still have this one."

Many of the books by Dr. Gupta were connected to the research papers. He explained, "We might work on a research paper and then realize there's so much more to say, we should make a book. For example, the book with Dr. Jie Chen on change point analysis evolved out of her thesis and new problems we uncovered."

A Lasting Contribution to Statistics

Dr. Gupta's research into and publication of the tables for Wilks' Lambda has had lasting significance. Dr. Tang explained this in a mathematical context:

> Dr. Gupta's work on Wilks' Lambda is part of the standard now. For example, if you go to use SAS (Statistics Analysis System), then Wilks' Lambda is the first one, along with Pillai's Trace. Because most of the test statistics depend on what we call the sample covariance matrix, Wilks' Lambda is called a determinant. While there are some people who use the trace or largest eigenvalue, smallest eigenvalue for the sample covariance matrix, Wilks' Lambda is always the first one. But there's a reason for it, because Wilks' Lambda is the likelihood-ratio test statistic and has very good statistical properties.

Dr. Tang went on to explain that the papers that Dr. Gupta and he wrote in 1984 are still actively referenced and cited as people continue to work on the topic. He explained, "One paper on the null hypothesis from 1984 had 96 citations. Another paper we did has 80. I think that's pretty good because in statistics or math, normally people are reluctant to cite works by others."

Cumulatively, Dr. Gupta's 529 publications listed on Researchgate.net show a total of 46,280 reads and 7,224 citations. To put this in perspective, the average number of publications by a full-time professor is about sixty over their career. Given the number of times the research has been read and cited, one could safely say his work has had an indelible impact on the field.

Of particular interest is that Dr. Gupta's research is the foundation for practical applications in use even today. One example is the research he did with Dr. Conradie in the 1980s. Dr. Gupta explained,

> We wrote a couple of papers researching distribution of different objects. It was some research that hadn't been done before. The conclusion is hard to describe in lay terms. But we derived the distributions in a form so that they can be computed."

A more dramatic outcome of this research was shared by Dr. Conradie, which is worth quoting in full.

> Some of the results that Professor Gupta and I derived jointly and independently in the early '80s were highly theoretical, highly mathematical stuff. And we both expressed our hope that it would stimulate future research to make it more attractive for practical purposes. In January 2008, twenty years after my visit to Bowling Green, I received an email from Professor Matthew McKay from the School of Engineering at the Hong Kong University of Science and Technology. The contents of the email said, in part,

> "Certainly, in the last decade, the remarkable classical random matrix theory results derived back in the '70s and

the '80s by researchers such as yourself, Professor Gupta, Constantine James... have now become extremely valuable and important for the design and analysis of modern MIMO (multiple input and multiple output) wireless communication systems, which is now probably the hottest topic in communications engineering. It is truly unbelievable the amount of people working in this area. As you may know, MIMO will actually form the cornerstone of the majority of future wireless technologies, such as next generation wireless, local area networks, cellular, mobile, etc. This is the Wi-Fi type of thing that we have today."

A simple definition of MIMO from the Internet explains, "MIMO exploits multi-paths to provide higher data throughput and increase in range and reliability without consuming extra radio frequency. It solves two of the toughest problems facing any wireless technology today: speed and range."

In 2012, Dr. Conradie received another email from Dr. Raymond Louis, who was in Australia. The email also highlighted the relevance of Dr. Gupta's and Dr. Conradie's results as applied to the MIMO technology. Dr. Conradie concluded,

> It is just a remarkable story of highly theoretical statistical results that we worked on in the '70s and '80s finding application in modern wireless technology twenty-plus years later. This story illustrates the impact of research in which Professor Gupta was one of the international leaders at that point in time. For any academic, experiencing that such theoretical research found an application many years later in a totally different field is very special. For me, it was just such a privilege to have met Professor Gupta, been part of his research network, and to have interacted with him over the years.

This particular incident has become part of Dr. Conradie's own teaching as he shares it with his students to encourage them in their research, telling the students, "In principle, if you have theoretical results, somewhere in the future, they might have a very nice application."

Dr. Gupta has always held that research needs to eventually have a practical impact. His daughter Alka shared that her father always tells the children and grandchildren, "Whatever you study, study hard, and be the best. If you want to do mathematics, don't just do mathematics. Do applied mathematics.'"

In short, without directly focusing on the application of his research, much of Dr. Gupta's work has had substantial influence in the real world, fulfilling his own mandate for others.

Keeping Current with the Field

A headline in the *New York Times* in 2012 reads, "What Are the Odds That Stats Would Be This Popular?" Indeed, what are the odds? The article goes on to explain that "Arcane statistical analysis, the business of making sense of our growing data mountains, has become high tech's hottest calling." Not only is the field of statistics popular, it is also constantly changing, unlike other areas of mathematics. As Dr. Gupta explained, "Statistics is very dynamic because it affects people." An article on the Harvard Department of Statistics website explains it poetically, "Indeed, the intellectual excitement, aesthetic beauty, and vast applicability are what make statistics as a scientific discipline flourish, from being considered much too narrow for Harvard half a century ago…".

Dr. Gupta's friend Dr. Jain commented that the field has changed so much, he sometimes has a hard time reading research papers now. He observed, "People in the field have become very specialized. Statistics has grown tremendously and continues to do so as new fields come in, such as the focus on data mining." Dr. Tang concurs with this, adding,

> Statistics is still based on a lot of traditional approaches, but the topics change. For example, nowadays, data analytics, data science, big data, or machine learning are the hot topics in statistics. In the good old days, we talked about the mean and the mean vector. But nowadays we try to find new methods, for example, to deal with the high dimensionality, a huge number of variables, and a lot of data. How do you do some kind of reduction in the number of variables or dimensions so that you don't lose a lot of information?

Keeping up with the evolving landscape is extremely important for Dr. Gupta, both as a researcher and a teacher:

> As a teacher, I always have to come up with new content. It is difficult at times, but you have to do it, because these PhD students have to be marketable, and unless they are up on the subject, their market value goes down. So you have to be up on the subject all the time.

This simple comment demonstrates why Dr. Gupta is a beloved teacher. He puts the needs of his students at the forefront of his activities.

In addition to the change necessitated by the demands of data, the field of statistics was dramatically impacted by the advent of new technology. As early as the 1960s, graduate students in statistics had to learn how to use the computer for data analysis. In fact, Dr. Gupta's work on Wilks' Lambda was pivotal, because the tables he created enabled computer programs to do the necessary calculations for Wilks' Lambda. Eventually, all aspects of Dr. Gupta's career became dependent on the use of computers.

> It affected one hundred percent of what we did. I had a research assistant who was good at computers. He was working for me the whole time, and he did a lot. Whatever you do research-wise in statistical research, you have to do some computations to support it. So, I wrote them and had the support of somebody else digitize them.

Despite the growing use of computers, Dr. Gupta still did most of the calculations by hand, explaining, "Most of the time I did them by hand. I was not good at computers, but the students were. That's why I said that the only qualifications the students should have was that they should have brains."

The evolving era of computers also meant that Dr. Gupta now could easily collaborate with others regardless of their location.

> It certainly changed how I worked with other people. I sent them emails after I knew what should be done or what could be done, and they did it. They were smart. The emails helped a lot, because I could reach Taiwan, China, and even Russia. I did not realize at the time how easy it was going to make things, but now I think that was a good thing.

How does a full-time professor, family man, and international lecturer stay current in the field? As time goes by, the books, papers, journals, and editors all change, so keeping up is a continuous process. Dr. Gupta's method is simple:

> You just read journals about what is being written, because the field changes; it is not static. The field keeps changing

and revealing the problems, and you have to think about them. The [professional organizations] were useful to me, because being a member means you get the journal that they publish, which keeps you up to date on what's happening.

In addition to providing journals, the association meetings are a means of connecting with others. Dr. Gupta mentioned that among others, the most useful organization has been the Institute of Mathematical Statistics (IMS), which he joined in 1966. IMS publishes multiple journals and holds a variety of conferences on different topics in the field, as does the American Statistical Association (ASA). In 1978, Dr. Gupta founded the Northwest Ohio chapter of the ASA in Bowling Green, attracting people from the university and from industry as well. This was one way of putting Bowling Green on the map, so to speak, in the world of statistics.

Attending seminars and conferences was a big part of Dr. Gupta's professional life. He explained, "Some of the information gets shared at conferences. I spent quite a bit of time at conferences, which is a freedom you get only in teaching. It's very difficult to get that freedom in other professions." He often met people who had mutual research interests, which would lead to joint papers, noting, "There was always somebody who I knew, and they came over to talk to me."

The purpose of going to a conference is two-fold: to learn something and to connect with others. Networking and seeing old friends are inevitable, especially for an established researcher, according to Dr. Conradie.

If you become like Professor Gupta—a very important, well-known person and academic researcher—you would have published a number of papers. Then you will be invited as a keynote or invited speaker to a conference, and you will share your research work of the last couple of years. When you are a young researcher, you will quite often go there and talk about the research you are busy with, and you hope for some interaction from the people in the audience that might contribute to further your research.

Dr. Jain described the meetings of the ASA that he and his friend Dr. Gupta attended:

The main meeting is where several thousand statisticians get together. It is usually about 30 percent from academia, about 25 percent from the government, and 25 percent from industry. I was in the industry part and Arjun was in the academic part. We have professional interaction with other statisticians from other parts of the country. And, of course, Arjun has given a lot of papers at these meetings.

Bringing the Heart Along

It was November, 1999, and the conference in Durban was already in full swing as everyone gathered in the small downstairs room for an informal gathering to meet and chat with each other. Off to the side of the noisy room, one of the most distinguished and renowned guests, a tall, white-haired Indian gentleman, was sitting on a chair, entertaining a twenty-month-old baby on his lap while her smiling mother looked on. Noticing this, some of the academics whispered to each other, "Can you believe that Professor Gupta is interacting with this little girl just like a grandfather would do?" Dr. Conradie, who shared this anecdote years later, commented,

I think for all of us, this showed the warm and gentle side of Professor Gupta. The little twenty-month-old girl graduated this year with a bachelor's degree at our University. We always spoke about that moment and said, "Oh, are you the one that sat on Professor Gupta's lap in 1999?" So that's something very small, but I think it is special.

Reading about Dr. Gupta's research activities can give one the impression of a man who lives in the rarefied atmosphere of the intellect all the time. Yet as the anecdote fully illustrates, Dr. Gupta has an innate quality of combining the head and heart, or his IQ and EQ. This incident in South Africa was only one of many that his friends and colleagues recounted. Dr. Jain related his own experiences of Dr. Gupta's genuine kindness:

> One ASA meeting was in Anaheim near Disneyland. For that meeting, I went alone with nobody else from my family. But his [Arjun Gupta's] whole family was there, and Meera had a cousin who lived about an hour away from Anaheim. Arjun made extra effort to have me tag along with the family to visit her cousin. Then another year, the meeting was in the Washington, DC area. Arjun had a relative who was working for the Indian embassy, so he took us to visit, and we had a nice dinner there.

Another anecdote from Dr. Conradie completes the picture:

> At our conference in Durban, he took time off to attend an annual general meeting (AGM) of our association, the South African Statistical Association. Now, the point is, it's not expected that your guest would attend your AGM. For most people, it is a boring affair. You just sit there, go

through financial and other reports, then some of the other people fight about this or that issue, and so forth. But he attended it. And my perception was that he was interested in the well-being of the association and the colleagues in South Africa. In fact, in his report on his visit to South Africa, he mentioned that he attended the AGM. To take time and attend an AGM from another association due to your interest in those people is very special, and we all experienced it as such.

~ CHAPTER 8 ~

AN INTERNATIONAL AMBASSADOR OF STATISTICS

By the time Dr. Gupta officially retired in 2015, he had completed over forty-seven years of continuous teaching, researching, and traveling. During that time, he gave lectures, taught classes, and presented papers in over twenty-three different countries and more than fifty cities. (See appendix E for a complete list.) Shorter trips were arranged around his teaching schedule. When travel was scheduled, Dr. Gupta had someone else take care of his classes. He explained, "Most of the time, I scheduled a test. Then I had teaching assistants who would help grade the tests." During the long hours of flying across oceans and continents, Dr. Gupta kept busy by reading research papers or revising the lectures he was going to give.

He often used his sabbatical for an extended stay in another country or school. On his last sabbatical before retirement, he taught at University of Michigan, only seventy miles from Bowling Green. He would go to Ann Arbor in the morning and come back in the evening. While the hour-plus drive was a relatively long commute, it was quite a short trip in comparison to the extensive journeys he was accustomed to taking.

Dr. Gupta's foreign travel embodies his unique spirit of an open mind and an open heart. He did not shy away from exploring unusual places, experiencing new cultures, and eating unfamiliar food, nor did he hold any preconceived notions about the qualifications of his prospective audience.

His extensive travels grew quite organically, beginning in 1978 as a visiting faculty member at Ohio State University in Columbus, Ohio, followed by a trip to Iraq in 1979, and an extended stay at the University of Campinas in Brazil in 1980. At both Ohio State and University of Campinas, he taught students, did research, and wrote some papers. Coming back to Bowling Green after Brazil, Dr. Gupta thought, "I want to do more of that kind of trip," because he clearly enjoyed it. Also, he shared, "There was more work that could be done in that way." Each time he traveled somewhere, he produced new research.

Being a global ambassador for statistics seemed to come naturally to Dr. Gupta. His daughters reported, "He was awesome at it, and he was invited everywhere. When he went to South Africa, he would enter, and it would be like some rock star walked in."

Dr. Gupta loved the adventure of traveling to different countries. However, traveling internationally from Bowling Green was nontrivial, often involving a trip to JFK in New York to catch transatlantic flights, then spending long hours in flight or in transit, which could be exhausting on the body. For example, he made a short visit to Kuwait, flying all the way there, enduring jet lag, standing up to teach, then within a few days, returning home again. It seems unfathomable, yet Dr. Gupta commented: "That wasn't easy, but I was younger, and I could do it." However, "younger" for the Kuwait trip was actually when he was in his sixties.

In the early years, when his three daughters were young, Dr. Gupta often traveled on his own. Later, when the girls were in college or elsewhere, Mrs. Gupta joined him from time to time. Occasionally, the whole family might be involved in a trip, such as to Vietnam, Thailand, or Japan. Some destinations were strategically planned to coincide with visits to his daughters, who also became world travelers themselves. Dr. Gupta's youngest daughter, Nisha, was living in Hanoi, Vietnam from 2009-2014, and Dr. Gupta and the family visited the area multiple times, attending meetings and conferences in both Vietnam and Thailand.

The diversity of countries he visited included some very economically and politically volatile locations. Reflecting on his global experiences, Dr. Gupta commented, "When I think of it now, it was quite dangerous going to all of those places—different food, different languages, different rules and regulations. But you just do it. You just do it."

Champion of Diversity

One could easily describe Dr. Gupta as a statistics missionary and a trailblazer who brought statistics to the far reaches of the world. With humility, he said, "Well, I tried my best." That, of course, is an understatement. Given the importance of statistics in the world, Dr. Gupta made a significant contribution, impacting many people. On a personal level, travel expanded his network of research collaborators. However, the trips also served the larger purpose of opening the field of statistics in many other countries. As a result, he also put BGSU on the map in the world of statistics, as many people, including his friend Dr. Vasudeva, have observed.

> I've heard that he has visited almost every country in the world where there is university education, to enlighten those people about his own field. He is a well-known scholar, and his visits brought a lot of students to Bowling Green State University who might not have even heard of Bowling Green except for his travels.

Colleague Dr. Jim Albert concurred:

Arjun is a great ambassador, because he's been everywhere around the world. I think Bowling Green is noted internationally because of Arjun. He's brought a lot of students. I think the best way to go about recruiting is to actually see people firsthand. He's done that and made connections. Just sending out an email or sending out a letter doesn't quite do it. Arjun was visiting places, giving talks, and talking about the program, which really makes a difference. So he was always a good ambassador.

Diversity and inclusion are crucial considerations in today's world, especially in the US. But years before people recognized the importance of giving everyone an equal opportunity, Dr. Gupta was doing just that. Bringing students from known and lesser known countries around the world to the middle of Ohio was groundbreaking in the 1980s and 1990s. As his daughter Alka put it,

> Somebody in the middle of Cameroon, for example, is not going to know about Bowling Green State University. They know it because Dad took the flag and planted it out there. It is a living example of how you level the playing field in the sciences and math.

Dr. Gupta's own assessment was,

> The trips were sometimes difficult in terms of living conditions or food, but that was where I met some students who came to work with me the next year. They chose the subject I was working on. For example, I met Amey in Ghana, and he came here and did his PhD thesis. And I met D. K. Nagar in India, and he came as a post-doc.

Dr. Gupta and D.K. Nagar later coauthored a book on their work together, *Matrix Variate Distributions*.

Dr. Gupta had a great ability to recognize and inspire quality students wherever he went, looking beyond their early education, which at times may have been insufficient. He was, of course, traveling to universities that had math departments. However, once the students came to the US to study, he would also invest five or seven years of his time as their advisor. Dr. Gupta explained,

> You start from the introduction of the basics, then go from there. Some of these students are innately mathematically inclined regardless of their early education. Supporting their study might not be easy. It's difficult but that is the understanding.

Dr. Gupta also traveled without prejudice and with an open mind. He has the skill of seeing the best in people, seeing the diamond in the rough, so to speak. He never assumed someone in an emerging country might not understand the subject. He attributes this attitude to his parents and his upbringing. He commented, "I remember my mother and father assumed that everyone was equal, but they needed some guidance."

Indra's Net

A metaphor in Eastern traditions, Indra's Net, explains the way everything is interconnected. It depicts a net the size of the universe, with every intersection of the net containing a jewel that reflects every other jewel. In many ways, Dr. Gupta's travels and personal connections evoke the intertwining and jewels of Indra's net.

For example, in 1983, Dr. Gupta was working on a topic in multivariate statistics, and a faculty member at the University of

Rajasthan, Jaipur, India, took an interest in it and invited Dr. Gupta to be a Visiting Fellow at the University. Dr. Gupta had a sabbatical leave coming up, so he accepted the invitation and went there for a semester, where he taught on multivariate tests. While there, he met a student named D. K. Nagar (Daya Krishna Nagar). According to Dr. Gupta, "When I lectured there, he asked me questions and followed up by coming to Bowling Green to do some work with me." After doing his post-doctorate work at BGSU, Nagar went to the University of Antioquia in the city of Medellín, Colombia. Many years later, in 1999, Dr. Nagar invited Dr. Gupta to Colombia, where Dr. Gupta gave a talk at the University in Medellín, Colombia. The symbolic representation might be something like:

Bowling Green—>Rajasthan —>
—>Bowling Green—>Colombia

Indra's net!

Dr. Gupta's travels were often at the invitation of someone he had worked with or who knew of his work. A researcher somewhere in the world would read Dr. Gupta's work and then might invite him to visit, lecture, or work on some new, related problems. Depending on the circumstances, Dr. Gupta might be a guest teacher at a university or a keynote speaker at a conference. Usually his most recently published research papers were the content for his speaking and teaching engagements.

In general, he lectured on some topic in multivariate theory, which he always enjoyed teaching, although when he was invited to teach at a university, the topic might be quite specific. Dr. Gupta said, "They invited me, they paid for my travel, and they told me what to do. This was not a problem, because whatever they asked, I had already taught either here or in India." Students in a foreign country noticeably expressed great curiosity on many subjects, creating lively interactions in the classroom. Dr. Gupta shared,

"When I traveled, they usually asked questions about me, and what I did before coming here."

Presenting papers at a conference was also enlivening and educational. As Dr. Gupta explained, "That's the whole purpose of the conference. You learn something new; you present your results, which are new; and you exchange ideas." Learning about new applications in the local countries and discovering how those countries used statistics was indispensable for keeping up with the field. The trips affected both Dr. Gupta's teaching and further research to a great extent. He explained, "I brought back to my students what I learned. Also, some joint research started to come out of these trips."

Dr. Gupta is definitely an adventurer, despite having what may appear as a serious demeanor. He gravitates toward new experiences, such as deciding to stop in Thailand on his way to India or making other impromptu travel plans. As his daughter Mita observed, "On one holiday to Italy, we just showed up in Rome with no hotel reservations, and he casually said, 'We'll just find one.'" This love of adventure not only permeated Dr. Gupta's life but left a lasting impression on the whole family. Mita shared.

> We all love to travel—all three of the daughters. I've been to eighty-two countries and my goal is over one hundred. It was so much fun and exciting to see him travel when we were growing up. He always liked seeing and experiencing something new.

Nisha similarly observed,

> By twelfth grade, I had traveled to at least fifteen countries; again, all due to Mom and Dad making sure we were

exposed to the world physically and immersed in the world mentally.

Traveling is so much a part of the Gupta tradition, that even the grandchildren "inherited it." Alka explained,

> In 2007, Dad gave a talk in Lisbon, Portugal, and the whole family went. It was my son's first trip at six months old. Mom and Dad were already there, and Mita, I, and little Arhaan flew to Lisbon. My husband Sharad joined us afterwards, and we all did some sightseeing, working the schedule around Dad's conference.

Ordinarily, one does not equate the life of a statistics professor with adventure. But Dr. Gupta was breaking barriers, going to places he might not have been able to visit as an individual, by traveling as "Dr. Gupta, Professor of Statistics." It was, in many ways, a courageous stance, given the political, culinary, and cultural differences he encountered. Somehow, his love of statistics gave him the energy to do it. As he said, "It wasn't always easy to do. Statistics was the only link that we had."

Traveling, by definition, means not only living out of your comfort zone but also eating out of your comfort zone. However, Dr. Gupta was raised in the traditional Indian culture, which instills the principle that one accepts whatever food is offered rather than insult the host by refusing it. Therefore, Dr. Gupta always did his best to be gracious and diplomatic about food. In China, Taiwan, and other parts of East Asia, the cuisine was particularly challenging for him, but he is amazingly flexible about food and willing to try exotic items. According to his daughter Mita, "As a guest, he tries whatever eyes or ears are put in front of him. In South Africa, he had a cow tongue. He's had frog's legs, and all types of offal. He has tried different things, but in the end, he desires basic North Indian food."

At the same time, travel wasn't always difficult or dangerous. Trinidad and Tobago, for example, were idyllic, with large, multigenerational, Indian-heritage populations. As Mita described, "They were on an island there for about a week, having Indian food, seeing the sights, and being treated kindly by everyone. Mom was also with him, which was great. They couldn't ask for more." Alka expanded on this.

> There were many great opportunities, and some that really fit with what they would want to do or a place that they wanted to be. He was invited to Rajasthan to give a talk. Spain is another one. And in 2007, we had a whole family holiday in Portugal planned around Dad's work engagements. Some of the other places, like Cameroon and Ghana in 1980s, were more challenging. But not all the trips were challenging.

The remaining sections present some of Dr. Gupta's extensive travels, illuminating his open-mindedness, spirit of adventure, and commitment to sharing statistics.

Early Adventures: Iraq, Brazil, Ghana

Dr. Gupta's early experiences represent testing the waters of international travel, even though perhaps he didn't view them that way at the time. Going to Iraq, Brazil, and Ghana moved Dr. Gupta well out of any familiar comfort zone regarding food, language, culture, and standard of living. Had these trips pushed him beyond his limits to cope, it is possible the subsequent years of circling the globe might not have taken place. Instead, Dr. Gupta discovered his own resilience and capacity to deal with whatever came his way and, in fact, enjoy the effort.

Iraq

"Students in Tehran Seize Embassy, 90 U.S. Hostages," read the headline on the morning of November 5, 1979. The year 1979 was a watershed year in the Middle East, particularly in Iran and neighboring Iraq. In 1979, the Iranian Revolution formally overthrew the US-backed Shah of Iran, creating an Islamic republic headed by Ayatollah Khomeini. Shouts of "Death to America" accompanied the storming of the US embassy in Iran in November, launching a year-long hostage crisis. Concurrently, in July 1979, Saddam Hussein took over as president of Iraq, claiming his power as a secular liberator. By the following year, Iran and Iraq were at war, fighting over territory and religion. In this highly charged, volatile situation, a US professor of the Hindu faith had the inconceivable courage to travel to the Islamic country of Iraq!

Despite some concerns from his family, during the winter break of December 1979, Dr. Gupta boarded a flight from New York to Baghdad. Dr. Gupta's willingness to make this trip in the midst of somewhat unstable conditions was based on the faith that all would be well, particularly because he trusted his host. On his arrival in Iraq, he was met by Dr. Sabri Al-Ani, a Purdue classmate who had returned to Iraq after graduation and was now a deputy administrator of education.

It had been twelve years since the two men had seen each other, and their greeting at the airport was warm, although Dr. Gupta managed to avoid the traditional Iraqi greeting of a kiss on both cheeks. He had jokingly told his family prior to leaving the US, "I'm ready to go, but nobody's going to kiss me." From the airport, the host and his guest headed into the city for an overnight stay. Dr. Gupta recollected, "We didn't see a lot going from the airport. We were there during the time of Saddam Hussein, so we were driven along the road that was pre-arranged."

The purpose of the Iraq visit was to meet students at the University of Basra, so the next day, his hosts drove Dr. Gupta to

Basra, about five hundred and fifty kilometers away. The six-hour car ride from Baghdad to Basra, a city that was later the scene of much destruction during the Iraq-Iran war, was not the safest, as the tensions were already quite high in the area. Dr. Gupta shared, "It was a long drive, but since Dr. Al-Ani was with me, it was okay." Dr. Gupta's summary of the trip was short and sweet:

> I was there for the graduate work rather than anything else. It certainly was very strange, but they treated us very well. I didn't know the local language, and their English was okay. Prof. Al-Ani did the translation in between. But it was okay.

Needless to say, when Dr. Gupta arrived home, everyone was quite relieved. At the same time, one could say, "Mission accomplished," as he had met the students, connected with his former classmate, and of course, had an adventure.

Brazil

In comparison to the Iraq trip, Brazil was almost tame and, perhaps more than any other trip thus far, whetted Dr. Gupta's appetite for more travel. The University of Campinas, commonly called Unicamp, is a top-ranked, public research university in the state of São Paulo, Brazil. It had been established in 1962, so when Dr. Gupta was invited there in 1980, it was comparatively new for a university. Describing the trip, Dr. Gupta shared,

> I went to Brazil for a semester and taught a course there. I didn't speak Portuguese, but the statistics classes were in English. There was also a United Nations center there. It was interesting because in Brazil, there was one person who worked with me, and we did some papers together. His name was Professor P. N. Rathi.

This first extended stay in a foreign culture brought out Dr. Gupta's adaptability. The climate was a bit challenging for him, since the summer there is quite hot and tropical. One compensation for the heat was the abundance of his favorite fruits —mangoes and guavas—that were next to impossible to get in Bowling Green. While Brazilian food was quite different from Indian food and mostly meat-based, he managed to work with what he was offered. Being a stalwart, open-minded traveler, he reported, "After some time there, I could say the food was good." In fact, after Purdue, he no longer considered himself a strict vegetarian, but rather an "opportunarian"—someone who eats what opportunity presents.

The language was challenging, the food was unusual, the students were different, and yet, Dr. Gupta concluded, "I enjoyed it." What was it that overshadowed the difficulties? His daughter Mita offered an insight: "For one thing, he was special and treated like a VIP. They took care of everything: housing, food, transportation." In addition, his work was well received, and he was able to do interesting research. After this semester-long trip, Dr. Gupta declared to his family, "I'd like to do more of these."

Dr. Gupta stayed connected to Brazil long after his first trip, in both his professional and his family life. On Friday night at home in Bowling Green, his family played Buraco, a card game he brought home from Brazil. Alka shared, "All of us learned to play Buraco, and then my son Arhaan learned it, so the grandchildren now are also playing it. The trip to Brazil was excellent in that way." In 1999, the entire family went to Brazil, spending New Year's Eve on Copacabana Beach in Rio.

Brazil was also the forerunner for similar trips over the years. Dr. Gupta took two trips to Mexico accompanied by Mrs. Gupta and one trip to Medellín, Colombia, to work with Dr. Nagar, as was already mentioned. While the trip to Colombia was a source of some anxiety for the family, Dr. Gupta felt that the statistical work there outweighed any perceived sense of danger.

Ghana and Togo

The third early trip is representative of Dr. Gupta's ability to travel to a highly unlikely place and return with a jewel from Indra's net—new, excellent students.

In 1982, after being a professor at BGSU for six years, Dr. Gupta earned a sabbatical, during which he worked as a senior consultant to the United Nations for a semester. In that capacity, he made a visit to Ghana. Dr. Gupta explained, "My connection came from a fellow who was in the administration there. He did some work in my area, and we connected by email."

Unlike Brazil, whose cuisine was different but tasty, Ghana was challenging. Dr. Gupta observed, "The food was very different, and I was hungry all the time. And the living conditions were not good at all." The University of Ghana, located in the capital city of Accra, had been established during the British rule in Ghana, which ended in 1957. The United Nations, Dr. Gupta's main sponsor, also had an office at the University of Ghana.

At the University, Dr. Gupta lectured on sampling theory and interacted with the students. During his semester there, two students stood out. One was Alphonse Amey, and the other was Sam Ofori-Nyarko. As Dr. Gupta shared, "They asked good questions and showed a keen interest in statistics and math." While their early education perhaps might not have matched an equivalent US education, Dr. Gupta supported their interest in pursuing PhDs in the US. He worked with BGSU to bring them to the US where they first started on lower-level courses to catch up with others. Eventually, they both earned their PhDs with Dr. Gupta as their advisor. Sam was at University of Wisconsin before returning to Ghana. Alphonse later became a professor, first at the University of the North in South Africa and then at Botswana International University of Science & Technology in Botswana.

Dr. Gupta's trip to Ghana was particularly memorable for his family, because he was almost completely out of reach for most of the trip. Telephoning from Ghana at that time was very difficult even from the central government office in the capital city. Mrs. Gupta related her experience.

> All three of the children were very young. I said to him, "As soon as you reach the destination, inform me." I didn't know what facilities they had. One day passed, then the second day, and the third day, but I didn't hear anything. And I said to myself, "My God, this fellow has gone, my three kids are here, and he has not done anything to reach us." I started crying. I called a friend and said, "I have not heard anything from him." I even called the United Nations in Ghana. I got no clue. Then one day finally, finally, he gave a call, just for two seconds, to say, "I'm okay." I got so mad. He's pretty good at keeping in touch, but that time was pretty hard. He literally didn't find any telephone.

During the Ghana trip, Dr. Gupta also visited Togo. The Togo trip is indicative of Dr. Gupta's pursuit of adventure. French-speaking Togo was not a very developed country at all, but there was somebody in the United Nations office who was going there to work, so Dr. Gupta went with him simply for the experience, commenting, "I didn't have to teach or give a talk."

South Africa 1999

"Black-maned lions framed against Kalahari dunes; powdery beaches lapped by two oceans; star-studded desert skies; jagged, lush mountains—this truly is a country of astounding diversity." This opening line from a South African travel site is not an exaggeration; scientists marvel at the biodiversity found in South Africa while paleoanthropologists have uncovered human civilization dating two million years old in the Kalahari Desert caves. It is no wonder that hearing the name "South Africa," the entire Gupta family becomes instantly animated, sharing stories about the country and the gracious reception they received there. South Africa is among the several locations that all the Guptas have visited, both as a family and individually.

The main point of contact in South Africa was Dr. Conradie, who first visited Bowling Green in 1986 for his sabbatical. Recalling that visit, Dr. Gupta observed, "He was good, and I offered him a regular position, but he wanted to go back. He was missing home too much." This is not surprising. Dr. Conradie clearly loves Stellenbosch: "I've been at Stellenbosch my whole career. Stellenbosch is one of the most beautiful places to live and to bring up children in South Africa, and it's also one of the best universities in South Africa, or in Africa, in fact." Dr. Conradie's six-and-a-half-month stay at Bowling Green coincided with Ohio's coldest winter and a record-breaking freeze, when temperatures plunged to minus 17°F in January! It was no wonder he opted to return to South Africa, where the record low for Stellenbosch is 68°F.

In 1999, Dr. Conradie was able to reciprocate the hospitality he received from Dr. and Mrs. Gupta. As president of the South African Statistical Association, he was instrumental in inviting Dr. Gupta to be a guest at the annual conference in Durban. Since Nelson Mandela had been elected only five years prior, Dr. Gupta explained, apartheid was still fresh in people's minds.

> I went to South Africa four or five years after apartheid had been abolished, and that was a big thing. In South Africa, I was considered colored, so some people asked Conradie, "Why are you inviting a colored man?"

Both Mrs. Gupta and Mita joined Dr. Gupta for what appeared to be a once-in-a-lifetime trip, though in fact, "once-in-a-lifetime" became multiple trips for each of the Gupta family members over a period of years. In fact, their youngest daughter and her family lived in South Africa from 2014 to 2019 during one of her international postings after Hanoi.

The first visit included a tour of the entire country, beginning in Stellenbosch near Cape Town. Dr. Conradie described the visit to the University of Stellenbosch.

> In the 1970s and '80s, we had a very strong multivariate statistics distribution theory group of people, and Dr. Troskie, my supervisor, was one of them. There were a number of people who Dr. Gupta knew through their publications. When he was here in 1999, he met most of them, and he was pretty well received. Due to his personality, he is just that type of person that is acceptable to everybody.

At Stellenbosch, Dr. Gupta mingled with colleagues in the Department of Statistics, offered a lecture, and enjoyed the sightseeing arranged by his hosts. One highlight was a visit to nearby Cape Town, which is home to hundreds of miles of nature reserve on Cape Peninsula, the geographical meeting place for the Indian and Atlantic Oceans.

Arriving in Durban for the conference, Dr. Gupta presented the opening keynote address, "Elliptically Contoured Models in Statistics," and offered a one-day workshop on "Statistical Analysis of the Multiple Change-Point Problem." It was at this conference that Mita and Mrs. Gupta observed what was described in this book's prologue—an enthusiastic reception for Dr. Gupta's work, where people lined up to get his autograph on a statistics book and meet him. "When he visited us in 1999," Dr. Conradie shared, "he was, at that point in time, a really world-renowned statistical researcher. If I remember correctly, he had quite a big audience at that conference for his two talks."

Durban is a beautiful port city with a substantial Indian population and an exquisite mile-long urban beach along the Indian Ocean. Hosted by Durban colleagues after the conference, the Guptas enjoyed another round of visiting, sightseeing, and familiar Indian food. A wonderful photo of Dr. Gupta on the beach in Durban captures his adventurous spirit—he can be seen risking getting his shoes wet as he reached down to pick up a shell very close to the incoming waves.

Dr. Conradie's former supervisor, Dr. Troskie, a professor at the University of Cape Town, had also come to Durban, joining a social dinner at a restaurant on the pier at the waterfront. Dr. Gupta said,

> In terms of intellectually challenging people, there was a fellow, Troskie, a South African guy. I had not met him before this trip, but his thesis at Stanford was very close to what I was doing. So I looked him up when I got to South Africa. During the workshop he asked me hard questions... in a good way. He challenged the thinking to the next level.

Leaving Durban, the Guptas traveled to the other major cities, including Bloemfontein, Johannesburg, Polokwane (formerly Pietersburg), and Pretoria, where they visited six different universities, met professors around the country, and also went sightseeing. For Dr. Gupta, "The most memorable moment was when they took me to Soweto. It was very moving." A stark reminder of apartheid, Soweto, in Johannesburg, still houses very impoverished blacks in either government housing or shanty towns. Nearby is also an "Asiatic town" that houses Indians who were also displaced during apartheid. Visiting a white university, an Indian university, and a black university, Mita recalled, "You could literally see the difference, as the facilities, amenities, and the upkeep changed, obviously because of funds and resources." It is interesting to note that in 1999, traveling to Soweto was still considered unsafe, so Mita and Mrs. Gupta were strongly advised not to join Dr. Gupta there. Later, in 2017, the entire Gupta family visited Soweto and were amazed at the modernization and cultural changes that twenty years had brought.

Dr. Gupta and his family also received a warm welcome from his former student Alphonse Kwame Adjei Amey, who was then teaching at the University of the North, a black college in Polokwane established under the apartheid regime's policy of separate, ethnically-based universities. The presence of Dr. Amey in South Africa is a vivid illustration of the network effect of Dr. Gupta's work. He met Amey in Ghana; Amey came to BGSU for his PhD and then returned to South Africa to continue the dissemination of statistical knowledge, particularly to those who might not have opportunities to encounter it. Dr. Amey accompanied the Guptas to the Kruger National Park, an exquisite and vast game reserve, teeming with wildlife, and full of archaeological history. Dr. Gupta summarized his South African trip:

> There was a lot of kindness extended there. In addition to Conradie, there was one professor, Odoom, who was very helpful. He was a professor of statistics who got his PhD in the States and went back to South Africa. I still have connections with him through email and wrote some papers with him, so the work goes on. Statistics was still pretty new in South Africa at that time, and it was very interesting. Now they have a statistical association also, and they're doing well. They have their journal, and the students are going out into the world beyond South Africa.

Asia

Dr. Gupta made a significant number of trips to East Asia, in part due to his wide professional network there, which included former students and BGSU visiting professors. Between 1997 and 2015, not counting family trips to India, he made fifteen trips to Asia, including Taiwan, Japan, Thailand, Vietnam, Malaysia, China, and Turkey:

1987:	Japan (Tokyo)
1997:	China
2000:	Taiwan
2002:	Japan (Tokyo, Kyoto)
2002:	China
2004:	Taiwan
2007:	Malaysia (Kuala Lumpur)
2008:	Turkey
2008:	Taiwan
2009:	Thailand (Bangkok)
2009:	Vietnam (Ho Chi Minh City and Hanoi)
2011:	Thailand (Bangkok)
2013:	Thailand
2015:	Thailand
2016:	Turkey

Taiwan

Flying into the island nation of Taiwan after a long, trans-Pacific journey over miles and miles of empty ocean, the relief from finally seeing land is accompanied by delight at the sight of the lush, mountainous island rising from the blue ocean. Landing at the modern airport in Taipei, travelers encounter a fascinating mixture of old and new, modern and traditional, temples and high-rise buildings, all comfortably mingled together. The sounds of Mandarin and the sights of unfamiliar Chinese characters on street signs can bewilder a novice traveler. However, by the year 2000, Dr. Gupta was no novice at all, and he navigated Taiwan with his usual undaunted spirit of exploration. As in other locations, he had excellent connections in Taiwan to make his stay quite comfortable. At the universities, the grad students in statistics knew English, removing a possible language barrier, and they were most eager to host their esteemed visitor.

Taiwan offered excellent opportunities for teaching and creating new connections with other researchers. On the first Taiwan trip in 2000, Dr. Gupta was a guest professor at the National Sun Yat-sen University, a research university located on a scenic beach in Kaohsiung Harbor in the south of the country. Dr. Gupta also taught at National Tsing-Hua University in Taipei that year, returning there again as a visiting research professor in 2004. Dr. Jen Tang, his former PhD student, was teaching PhD classes in Taipei at that time while on four-year family emergency leave from Purdue. While Dr. Tang regretted not being able to spend a significant time hosting Dr. Gupta, he was of great help orienting his former professor to the country.

During 2008, Mrs. Gupta joined her husband for two months during a longer stay at the newly founded National University of Kaohsiung. Dr. Mong-Na Lo Huang, a Purdue graduate, headed the Department of Statistics there, and he and his wife helped the Guptas navigate the complexities of daily life in Taiwan. Dr. Gupta enjoyed some of the local food, such as his beloved mangos and lychees, but overall, he found Taiwanese food an adventure at best. However, he and Mrs. Gupta were housed in the University accommodations, which afforded them the possibility of home-cooked food, making their stay more comfortable. The apartment, located on a bit of a slope, also provided an excuse for exercise as Dr. Gupta walked to and from the campus.

Nisha recalls visiting her father in Taiwan and exploring the night markets, where she convinced her father to try all the street-food she found so enticing. However, she said, her father drew the line at the coagulated duck blood soup. Dr. Gupta was quite impressed by Taiwan.

> Taiwan was an interesting country. It was really a very advanced country. The lodgings were comfortable. They had everything that I thought we had here. They had a modern digital infrastructure, their own journals, and had done quite some work in statistics. I was impressed. Language was a major hurdle for anybody who was not a statistics major. But they would try to be friendly. Their personalities were generally serious, and they were very hardworking. People worked six days a week, sometimes more.

When asked what he regarded as his most important contribution in Taiwan, Dr. Gupta replied with typical humility.

> Well, the only thing I can say is my lectures were useful to them. I lectured on distribution theory. It was a very advanced country, not a developing country, and I enjoyed going there.

> Overall, Taiwan was an enjoyable place to work and experience an East Asian culture. Mrs. Gupta shared, "I enjoyed Taiwan. I was there for two months. People are so nice and friendly, and you're absolutely free to do anything and go any where. And they were very, very nice, even though the language was a little problem."

Japan

Japan is a beautiful country with an aesthetic sensibility quite distinct from other East Asian countries. Visiting Japan, one glimpses many common roots with the ancient Chinese culture, from food and art to architecture and gardens. The enchanting beauty was not lost on Dr. Gupta, who traveled there twice on business in 1987 and 2002. Dr. Gupta's wife and his three teenage daughters accompanied him on the first trip, whetting their appetite for more visits.

The trips to Japan vividly illustrate the power of connections formed at Bowling Green, as Dr. Gupta had no less than three significant hosts during his visits: Dr. Konishi, Dr. Hisao Nagao, and Dr. Sugiyama, all of whom had come to BGSU at one time or another to work directly with Dr. Gupta.

The focus of the 1987 visit was the Biennial International Statistics Institute World Statistics Conference, which drew luminaries of statistics including Professor C. R. Rao. There was a heightened sense of excitement around the opening of the conference as Prince Akihito gave the welcoming address. The very next year, Akihito became emperor when his father Hirohito died. The whole family had a chance to attend the opening ceremonies, after which Dr. and Mrs. Gupta went to the conference celebratory party while Mita, Alka, and Nisha enjoyed exploring the area on their own.

Japan's economy was at its most expansive stage at that time, and prices for goods were at their highest, as Mita relayed in this anecdote.

> We were in Ginza, and Nisha wanted to buy a peach. It was five dollars for one peach. It was wrapped really nicely, like jewelry, but still not worth five dollars!

Mrs. Gupta's brother, Dhruveshwar Nath, was living in Tokyo during the Gupta's visit, creating an excellent opportunity for the family to get to know Japan from the inside. He showed the Gupta family around Tokyo, then they took a bullet train to Kyoto, passing the majestic Mount Fuji along the way. Dr. Gupta gave a lecture in Kyoto, and the family visited a number of the ancient Shinto and Buddhist temples there.

In 2002, Dr. Gupta and Mrs. Gupta once again enjoyed the gracious hosting and unique culture of Japan. This second trip to Japan was at the invitation of Dr. Sadanori Konishi, Professor of Mathematics at Kyushu University, whose research interests include multivariate analysis. In the mid-1980's, Dr. Konishi, his wife, and twin daughters had stayed in Bowling Green, where he was a visiting professor. Dr. Konishi was delighted to be able to reciprocate the hospitality by welcoming Dr. and Mrs. Gupta to Japan. Dr. Gupta and Dr. Konishi published several research papers as a result of the visit.

Unlike his experience with some other types of food, the Japanese cuisine was enjoyable for Dr. Gupta. His hosts went out of their way to honor his vegetarian preferences, taking their guests, for example, to a restaurant that served only variations of different tofu dishes. Dr. Gupta observed,

> As far as the food, they made it quite tasty for us, and we liked it. They made it all vegetarian. We also had sukiyaki, which I ate. Sukiyaki is a dish they cook up in front of you. There's no equivalent in Indian food, where you sit and you all cook the same thing together on the table.

During this second trip, Dr. Gupta reconnected with Dr. Nagao, who had also visited Bowling Green a few years earlier. Dr. Nagao escorted Dr. and Mrs. Gupta on some sightseeing trips around Tokyo, as well as a visit to the Hiroshima Peace Memorial Museum in Hiroshima. Dr. Gupta recollected,

> Visiting the Hiroshima museum was interesting and eye-opening. For example, there was a Japanese soldier that we [Americans] had shot, and his photo was there. He was a

hero there. It was quite emotional and a little uncomfortable to be there.

A third connection in Japan was Dr. Sugiyama. Dr. Gupta explained, "Dr. Sugiyama was a Japanese statistician who came to do some work with me. He had read my papers and contacted me. He was very interested in multivariate theory. When I went to Japan, I visited with him."

Although Dr. Gupta is very appreciative of the culture and beauty of Japan, his focus always returns to statistics when he assesses a trip.

In addition to the country being advanced, they were also advanced in statistics. They had their own journal, and they had a Department of Statistics. Not every country or school has one. Some of the people in that department made big contributions to statistics, such as Dr. Okamoto. He saw the problems, which a professor at Stanford, Dr. Anderson, had proposed. He [Okamoto] solved the problems, and then he was able to write papers on them.

Thailand

One location in East Asia that attracted Dr. Gupta's attention was Thailand, and over a period of a few years, he made multiple visits there. He first explored Thailand as a tourist after his graduation from Purdue. "In 1967, I was going to India," he explained, "so I asked my airline to give me a ticket that went through Thailand." Apparently, Dr. Gupta's appetite for adventure was embedded in his life even as a young man.

In 2009, he made his first trip in a professional capacity, accompanied by Mrs. Gupta.

In 2009, there was a student from the University named Dr. Wong who knew me and invited me to Thailand. My wife especially liked the Buddhist temples and the peaceful atmosphere. Buddhism permeates everywhere in Thailand and has an influence. There's a certain attitude of nonviolence, much like Gandhi. In Thailand, we did some work and published a paper, but there was a shortage of books. I went back two more times because of the joint work.

During his visit in 2009, he taught workshops at King Mongkut University in Bangkok on "Asymmetric Models in Statistics" and "Change-Point Analysis." At that time, Nisha was living in Hanoi, Vietnam, and she and her husband, Patrick, joined her parents in their travels that year and again in 2011. This made additional adventures possible. Nisha shared,

> On one of the visits, I met them in Bangkok, and we flew to Siem Reap, Cambodia, to see Angkor Wat, where we met up with my husband. Then from there, we flew to Hanoi. On another visit of Dad's to Bangkok, we celebrated Thanksgiving (Patrick, Dad, & I) at a Pizza Hut opposite Thammasat University campus, hoping for turkey pepperoni. Dad wasn't in want for food though, as he had visibly packed on a few pounds from all the yummy rice and noodles at the University canteen.

In 2011, Dr. Gupta returned to give the keynote address at the International Conference on Mathematics, Statistics, and Applications in Bangkok, Thailand. His talk was entitled "Estimation and Model Selection Based on MPS for Multivariate Skew Normal Family." At Thammasat University in November 2013, Dr. Gupta offered a short course on the theory of multivariate statistics. Another host in Thailand was Dr. Wang, and over the period from 2008 to 2019, Dr. Gupta and Dr. Wang collaborated on and published at least five papers.

Vietnam

To many Westerners, travel to Vietnam may feel somewhat exotic, as it maintains its cultural uniqueness in most villages and remote areas. However, it is yet another location that Dr. Gupta handled with apparent ease. In 2013, he and his family flew to Hanoi where his daughter Nisha and her husband were posted. After offering a lecture at the University of Hanoi, they all flew down to Ho Chi Minh City where Dr. Gupta presented a talk at the Ho Chi Minh City International University. He also gave a keynote address in Ho Chi Minh City at a conference on "Statistics and Its Interaction with other Disciplines." When asked if a trip to a country like Vietnam felt unusual, Dr. Gupta replied:

> Talking about statistics is never unusual for me. Statistics is new work everywhere. So, for me it was easier, and for them it was new. For me, the trip to Vietnam was really interesting.

China

The only country in the world that rivals India for size and longevity of culture is China. While in many ways, you cannot find two more differing countries, there are also similarities, and often travelers to both countries might compare the two, sometimes favoring one or the other. Mrs. Gupta, who accompanied Dr. Gupta on two trips to China, had this observation:

> I think that their family values are basically the same as the Indians. You may not follow their language, but they are

basically like the Indians in the way they respect the elders in the family and how they treat the children. The elderly person walks in the room, and a small child or somebody will get up from the chair to give it to the elderly person. Same as in India.

The year 1997 marked a significant opening in the US-China relationship—the heads of state exchanged visits that year in an attempt to portray harmony. One expression of that harmony was the Ambassadorial Delegation (Statistics) lecture and tour to China in October 1997, in which Dr. Gupta participated. It was a delegation of about fifteen people in mathematics and statistics. As part of the visit, a conference was held at the Beijing University of Technology. Everyone at the conference spoke English, which is a required language of study for any Chinese professional.

Dr. Gupta's close friend from Purdue, Dr. Jain, also attended the conference, accompanied by his family, providing an opportunity for the families to tour around while the two statisticians took part in the conference. A few years later, Dr. Gupta returned to China with Mrs. Gupta, and he finally visited the Great Wall as well as other well-known tourist sites. Mrs. Gupta described their experience.

> I went to China twice. One time, there was a meeting in Beijing, and the second time, a student invited him in Shanghai. That was really, really enjoyable. They made sure we were comfortable in the hotel room and that we saw all the highlights of the city and elsewhere. We went to the Great Wall, Tiananmen Square, the Forbidden City, and the old Summer Palace. In Shanghai, we went to the Shanghai tower and the Bund.

She continued,

You can talk about anything you want, but don't speak ill of the Chinese leaders. Of course, the parts of the city we were in were very clean, but they showed us only that part of the city. I don't think the rest of the country is like that.

While China's village system manages to house and feed people, it has its fair share of poverty in the cities, similar to any other large country. At the same time, it has produced excellent researchers in the field of statistics, which was of prime importance for Dr. Gupta's visits.

Turkey

In 2005, Dr. Gupta's first trip to Turkey, geographically part of both Europe and Asia, was like a dress rehearsal for a very successful second visit in June 2008, when Dr. Gupta was the guest of honor at a conference in Kayseri, Turkey. The conference, entitled "Multivariate Statistical Modeling and High Dimensional Data Mining," was organized by Dr. Gupta's former student at University of Michigan, Dr. Hamparsum Bozdogan, who was originally from Kayseri. The conference took place just prior to Dr. Gupta's seventieth birthday, providing a convenient way to combine a professional gathering with a celebratory occasion of sightseeing. It is noteworthy that even at the age of seventy, Dr. Gupta's interest in adventure and traveling had not diminished a bit.

At the conference, Dr. Gupta gave a keynote address on "Ubiquitous Gaussian Distribution and Modeling Skewness." According to Dr. Conradie, who was also there, "The conference was a special occasion, where Professor Gupta was honored. If I remember correctly, at one of the social events of the conference dinner, there was a toast to Professor Gupta." Dr. Conradie also described the post-conference experience:

Kayseri is in the area of Cappadocia, and it is a very special and interesting place, especially the unique rock formations, the so-called fairy-chimneys, and the Göreme National Park (a UNESCO heritage site). A special moment for us was when we were together in a bus to visit all these places. That day we had a wonderful situation. We had the privilege, while walking through all of these places and riding through the countryside, to spend some quality time with Professor Gupta. I always remember that.

Dr. Gupta also has fond memories of Turkey, particularly due to the sincere and warm hospitality shown by his former student. He returned a third time in May 2016, invited by Dr. Bozdogan to give the keynote address at the International Conference on Information Complexity and Statistical Modeling in High Dimensions with Applications (IC-SMHD-2016). As part of that conference, his former student Dr. Bozdogan was honored with a *Festschrift*, a volume of writings to which Dr. Gupta had contributed.

The Middle East

Reading the past thousand years of India's history, one clearly sees evidence that Hindus and Muslims previously lived together harmoniously and respectfully, even occasionally borrowing traditions and customs from each other. They fought together against the British in 1857. When Arjun Gupta was growing up in Purkazi, Muslims and Hindus lived together. He shared, "At high school, there was no distinction between Muslims and Hindus. Some of the teachers were Muslims. Nobody tried to convert anybody else to their point of view."

However, during the struggle for India's independence in the late 1940s, any sense of harmony was completely dissolved as the British Government promoted animosity between the Hindus and

Muslims. Dr. Gupta knew the effects of this firsthand when his elder sister fled Pakistan. Yet without hesitation, he accepted offers to visit and teach in Islamic countries—Iraq, Kuwait, Saudi Arabia, Egypt, and Sudan—embodying a religious tolerance he learned from his parents. His friend Dr. Vasudeva observed,

> He's extraordinarily open, because he's traveled to, for example, the Asian countries where Buddhism is popular and to the Islamic countries. I think most of us come to realize, after being exposed to diverse cultures and religious practices, like various aspects of Christianity, Islam, or Buddhism, that it is a matter of faith. Your faith is personal, and it doesn't have to become a barrier to social interaction with people. In fact, it is an opportunity to learn from other people. This is how Dr. Gupta traveled.

When asked how he went as a Hindu into a Muslim country, Dr. Gupta's answer is very telling:

> I didn't go as a Hindu or an Indian. I went as a statistician. And there was always somebody who knew me or my work, and they took care of me. I never had an inner sense of discomfort, except in one place. In Saudi Arabia, my wife and I wanted to enter a restaurant, and they said, "women not allowed." Other than that, we were always taken care of.

Religious and racial discrimination were not in the foreground of his mind, and he had no difficulty complying with requests such as, "Don't talk about the government, and don't talk about the Israeli problem." As he said, "I talked about statistics."

In 2010, accompanied by Mrs. Gupta, Dr. Gupta traveled to Cairo, Egypt, to attend the International Conference on Modeling,

Simulation, and Control. He gave the keynote address, "Estimation and Model Identification Based on MPS for MSN Family." Then he and Mrs. Gupta did some sightseeing, which he said "was fantastic, full of history and many monuments." Since Egypt is more secular and perhaps more open than some of the other Middle Eastern countries, the trip to Cairo was rewarding from both the statistical and cultural perspectives. In the 1980s, Dr. Gupta also paid a visit to Sudan, which was, like Egypt, a little more welcoming to him than perhaps the other countries in the area.

Saudi Arabia was quite a different experience. In 2012, Dr. Gupta was the guest at King Fahd University of Petroleum and Minerals, a sprawling institute named for a former king of Saudi Arabia. Again Mrs. Gupta accompanied him, but this time some surprises awaited them. There were serious restrictions placed on Mrs. Gupta as a woman, and she was not even allowed on the University premises. Furthermore, she and another faculty wife were not permitted to go out alone or leave their guest house on their own. They ate their meals delivered to their hotel room. Describing this uncomfortable experience, Mrs. Gupta said,

> Saudi Arabia is beautiful, but I just wanted to get out of there as soon as possible. Maybe it was also the mindset, hearing things like, "you can't go there without covering your head, or they'll arrest you." I just barely walked outside. We were there for one week. I said to Arjun, "Can you go after three days?" And he said, "I need to finish my lectures. I can't just leave in the middle."

> We were in Dhahran with the gas company there, Amoco. We were right on their campus, which had a huge mall. One evening, I was walking with my husband at the mall,

and a woman came close, pointing up and down at me, saying "*Abaya, Abaya.* Where is your abaya?" [*Ed. Note*: An abaya is an outer garment worn over street clothes by women in Saudi Arabia, whenever a woman leaves her home.] I had decent clothing on. I had a scarf here, and I was wearing a long tunic with pants. I said, "Oh, my gosh, let's take a taxi, and go back to our room." There was a tea store right inside the mall. I had said, "Let's have a cup of tea." But after what that lady did, I said, "I'm not going to sit here." I just felt so unsafe. I felt like, "Are they going to come and arrest me?"

In the end, the trip was a reminder that travelers cannot be completely protected from deeply held cultural expectations. Nevertheless, it was the first and only time Dr. Gupta encountered this kind of issue in all of his travels, which is a testimony to the goodness of his mission and the people who helped make his travels possible.

Home Again

In the same way that the roots of a tree enable it to spread its branches wide, Dr. Gupta's family provided the roots for the enormous scope of his world travels. Dr. Gupta's journeys started and ended from his home in Bowling Green, and during each trip, he kept in touch with his family regularly, except during that one incident in Ghana where there was no phone!

At home, the girls would locate the destination of their father's upcoming trips on their globe, raising their awareness of the world beyond their own town and stimulating their own interest in world travel. In anticipation of their father's trips, the children would read about the country in the volumes of the *Encyclopedia Britannica* that were always kept within easy reach in the family home, since this was clearly pre-Google days. The well-worn (and perhaps now inaccurate) globe is still in the Gupta home and is a family heirloom for the grandchildren.

On the day of Dr. Gupta's return to the US, everyone would eagerly pile into the car and head for another airport adventure. Mrs. Gupta made sure they left early enough to have time for treats at the airport, such as their favorite mashed potatoes. Then they waited for him to emerge from customs. The girls would compete for who could see their dad's shoes first as he descended on the escalator to the baggage claim, where they were eagerly waiting. The heartfelt welcome Dr. Gupta received from his family was a stark contrast to his first arrival in America decades earlier, when there was no one to greet him.

The car ride home from the airport was filled with stories of his adventures as well as a sense of anticipation, since his daughters always knew they and their mother would get some special souvenirs from his trip. At home, their father would empty his suitcase and hand out the treasured gifts. In Basra, Iraq, Dr. Gupta bought a pearl necklace for his wife as well as some loose pearls, which were later put in Alka's wedding necklace and in a necklace for Mita. These were particularly precious, since the famous pearls, or *moti*, from Basra were widely known in India. Another treasure was a beautiful amethyst from his trip to Brazil. Alka shared, "I know what amethyst is because of that trip. He also bought a beautiful blue plaque ornate with butterfly wings that is still in our house." And Nisha treasures the onyx sculpture of grapes that he brought home from one trip to Brazil. The gifts for the family created lasting impressions on the girls about the beauty of the world as well as instilling in them the practice of generosity that their father consistently demonstrated.

Alka and Mita also had a collection of dolls from different countries that their father visited, including a doll of a man with a turban that was brought back from Iraq. Dr. Gupta often involved his host in the effort to find a suitable doll that represented the country. It is easy to imagine the delight it brought his hosts to help their renowned and honored guest do something so ordinary and human. This innocent request for help vividly illustrates Dr. Gupta's warm-hearted way of connecting with people, and the way he kept his family with him no matter where he traveled.

As the children grew older, more trips involved the whole family, eventually including the next generation (grandchildren). Even when the girls were living on their own, travel became a unifying theme for everyone as they managed to meet in a variety of locales, including East Asia, Brazil, South Africa, and of course, India.

In short, Dr. Gupta's travels started from home, ended at home, and sometimes, he simply brought everyone from home with him. Whether traveling on his own or with his family, Dr. Gupta's international reach made education and statistics accessible to those who might not have had access, changing people's lives dramatically.

- Photo Gallery -
Part III: The World

Iraq, 1979. Basra University colleague (left) and Dr. Gupta (right)

Dr. Gupta in Sudan with his transportation, 1982

Part III: The World

Stellenbosch, South Africa: Mrs. Gupta, (Left), Dr. Gupta (middle) with their host Dr. Conradi, 1999

Dr. Gupta on the beach at Durban, South Africa, 1999

Prof. Troskie (left) and Dr. Gupta (right) in Capetown, South Africa. 1999

Durban, South Africa, Dr. Gupta (right) with host professors

Visiting the Durban Botanic Gardens in Durban,
South Africa with hosts, 1999

Dr. Gupta and Professor Delia North at
Howard College, Durban, South Africa 1999

Enjoying mangos in Taiwan

Dr. Gupta (far right), next to Professor and Mrs. Huang, and two students (left) in Kaohsiung, Taiwan

Part III: The World

Dr. Gupta with students in Kaohsiung, Taiwan

Enjoying the gardens of Taiwan

Visiting the Hiroshima Peace Memorial Museum in Hiroshima, Japan. 2003

Mrs. Konishi and one of her twin daughters (left) having dinner with Dr. and Mrs. Gupta in Japan

With Professor Konishi in Bowling Green

At a statistical meeting in Tokyo, 1987.
From Far Right to left: Prof. Segura, Prof. Konishi, Dr. Gupta, Mita,
Mrs. Gupta, C. R. Rao. Far Left: Mrs. Konishi, Prof. Sugiyama.
Front: Prof. Konishi's twin daughters,

Mrs. Gupta and Professor Nagao in Japanese Gardens in Osaka

Mita, Meera Gupta, Arjun Gupta, Nisha, and Alka,
Hoi An, Vietnam, 2009

Part III: The World

Doll collection

Arjun Gupta with his wife, children, and grandchildren, December, 2018

The future looks bright! Arjun and Meera Gupta with grandchildren Arhaan, Saamik, and Anika.

~ CHAPTER NINE ~

REFLECTIONS

The greatest wealth is to appreciate what we have and what we are.
                                                        Arjun Gupta

This maxim that Dr. Gupta passed down to his children as guidance for their life encapsulates a great deal about his own way of living. He lives with appreciation for what he has, and he clearly knows who he is, moving about the world with an unshakable inner confidence. As this biography concludes, it is relevant to reflect and ask, "Who is Arjun Gupta?" Perhaps the voices of his friends and family as well as his own voice can summarize the answer.

Friends for Life

The line between friends and family is sometimes hard to discern in Dr. Gupta's world. Traditionally in India, children refer to their elders by the honorific title *Uncle* or *Auntie*, so the line gets even more blurred. Thus, the Guptas have a significant number of close, lifelong friends who feel like family, and family members who are close friends. Arjun Gupta's friends often use similar terms to describe his qualities and character. The noticeable quality and depth of his friendship was described by his friend Dr. Bansal.

I have not developed a friendship at that level in this country with anybody. Arjun is like a brother. Despite all my friends in town here and my colleagues, a real sense of older brother exists only with him. The depth of political, economic, and news discussions we had is sincerely a joy. I couldn't have these with any other friends, because none of them are as well-versed or have the depth of knowledge in different areas that he does. Maybe that's why I felt the first time I shook hands with him that I had met my older brother.

Loyalty and caring are also significant attributes of Arjun Gupta's friendship. A good friend Dr. Jain described one incident that took place when he worked for Bell Labs in New Jersey.

One time, my wife took me for dinner to an Indian restaurant in our neighborhood. I was just going for the regular dinner, but it turned out it was a surprise sixtieth birthday party for me. I didn't know that. I just went for a regular dinner, and there he was. He and Meera took the trouble of driving all day [from Ohio to New Jersey] and coming just for that sixtieth birthday party. He's a very loyal friend and goes out of his way for people that he cares about.

Many friends, family, and colleagues use the word *inquisitive* to capture the sincere interest that Dr. Gupta takes in others. Dr. Conradie explained,

In the way that he steered the conversation, the way he asked questions without ever overstepping your personal space, I always got the feeling that he was interested in your opinion and interested in you as a person. He asked questions about you, about your family, about where you come from. He asked, "How do you do this and that in your country?"

Showing interest in others is the ultimate form of empathy, as people generally have a need to be seen, heard, and understood. Dr. Gupta extends this gift to most everyone he meets. His nephew Vishwanath Prakash said:

> He would ask me questions and not just merely as a point of conversation. He takes an interest and asks me a lot about my job and my experiences. So that's one thing that comes across to me—his mind has always been inquisitive, and he wants to know, he wants to absorb and imbibe anything and everything around him. That is a great quality of his, which I have always admired. He would always keep asking, take interest, and go into more and more details.

Dr. Gupta's niece Jyoti explained, "He's one of those people who is interested in everyone in the family." She continued:

> He wants to know, "Okay, what are you doing now, and why are you doing it?" He's very family oriented, not just to his nuclear unit but to his extended family. When he hears something, he'll write it down, and he'll write down the relationships. Then the next time you would call, he would remember the spouse's name, your kid's name, or what happened when.

Jyoti concluded with these thoughts about her uncle:

> I would say he is highly intelligent. He has both high IQ and high EQ. I think he is one of the few people who has

both. He would really go deep into a subject, and he would be very focused, very intense about it. He's a very interested person and a very loving person.

His Greatest Success

Countless world travels, research papers, successful students, lectures…All these areas of success make it easy to overlook the most obvious success: Dr. Gupta's family. His family was the anchor for his wide-ranging activities, and they have been a source of nourishment throughout his life. As he shared, "I really enjoyed being with them. They gave me a reason to leave the office." When asked what he was most grateful for, Dr. Gupta responded,

> The answer is very difficult to give, but I think my children are my best, and what I am most grateful for. My primary goal was education, and they did very well in their chosen fields, and I'm proud of it. It is difficult to pick one thing out, but in general, they have done very well, and I'm satisfied.

There is little wonder that he is satisfied. Alka, Mita, and Nisha are not only well-educated, professionally successful, and independent, but they also imbibed personal qualities from their parents that are universally admired by all who know them. In interview after interview, colleagues and friends would say, "If you want to know what kind of person Dr. Gupta is, look at his children." Dr. Gupta's nephew Vishwanath Prakash put it this way:

> The ultimate reflection is the children. To meet Alka for even five minutes wipes away so much. She is such a positive person, and it is such a pleasure, always, to meet Alka, Mita, and Nisha. It is beautiful to see his children. It's the children who reflect what the parents have been.

Mrs. Gupta shared, "All his life, he worked hard, and he's devoted to the family. This is a very family-oriented man, and he will do anything for the girls, anything. And now the grandkids. Those are the stars of his eyes, I guess." She went on to describe their daughters:

> They care a lot for us. I mean, a snap of the finger, and they will come back to help. I guess they just learned that you have to be together, you have to give each other the support. Whatever medical thing is happening, is happening. But you've got to give support to each other. Being young, growing up, they could argue with each other; all these three sisters could fight with each other. But when the time comes, they know. That's what counts.

As a close family friend, Dr. Bansal had a chance to observe the family over many years. He commented on Arjun's demeanor as a person, saying:

> The kids understood what was expected. He and his wife made a great team to raise a family. The kids never had any issues about, "Mom is this tough" or "Dad is that tough." If that were the issue, at least once in a while you'd hear from the kids, and none of the kids complained. I think Arjun definitely held expectations for his daughters, and they've all lived up to it very well.

Each daughter's life is unique, yet they have been clearly influenced by their parents and their upbringing. They are all confident and self-assured women, with a sense of empowerment. Mita shared what contributed to this.

My entire formative life was in Bowling Green. We didn't move around. I had all the same friends. It was just so secure. You take it for granted when you're growing up; you don't realize until you talk to other people how valuable it is.

Alka continued this theme, saying:

We were allowed to do anything we wanted and always with a sense of empowerment and responsibility. He taught us that you are always empowered with a way. And if not, then so be it. You have all the resources, so figure it out.

And Nisha independently echoed the same theme, saying,

What I appreciate most in retrospect is that no opportunity was ever denied. If I wanted to do it, then with no hesitation, I got to try it, and often realized I didn't want to do it—for example, trombone and Russian classes.

For Dr. Gupta, every moment was a teaching moment for his children, whether at home or abroad. Nisha described her experience when her father brought the whole family along on one of his world trips:

On those childhood visits to other countries, my dad would often ask me to plan the itinerary or go pay the restaurant bill, to which I would respond, "But I don't know the language," or "I can't convert the currency." To which he would then respond, "Then go learn," or "*akal badi ki bhains*" in Hindi, which literally translates to, "Which is bigger, your brain or a buffalo?" To my chagrin, I now do the exact same thing with our two young children, and I know exactly why.

Dr. Gupta has instilled impeccable values and traits in his daughters. Four traits in particular stand out as a direct reflection of their father: a passion for global exploration, the desire to excel in their fields, a commitment to working hard, and a deeply held value of education.

Dr. Gupta's daughters have all made travel part of their lives, and they are naturally world citizens, comfortable anywhere. At any one time, each of them might be on a different continent. One week during the creation of this book, Nisha was in Bishkek, Kyrgyzstan; Mita was in Spain; and Alka was boarding a plane to fly from California to Ohio. Their lives have unfolded in a way that uncannily parallels their father's inclination for people and travel.

Alka is at the forefront of corporate governance and financial technology. She is known for her her ability to create connections between people, a skill she attributes to observing the same talent in her dad. Mita has a thriving career in building businesses globally—living the notion of "knowledge is power"—quite often being one of the smartest in the room. Nisha's career as a health scientist in the U.S. Foreign Service takes her all over the world for years at a time. She is now raising her children with the same or higher level of international exposure that she had when she was raised.

Excelling is part of the fabric of all three daughters, who exemplify their father's commitment to excellence. When asked about the influence of her father on her education, Nisha shared, "Influence is an understatement—more a mandate: do something useful, and whatever you choose, be the best in your field. Full stop." Similarly, Mita learned from her father, "Go to the top, go to the best institution, and excel. Study hard, work hard, do your absolute best." Alka recalled that once in an interview, someone asked her, "Where do you get your drive? It's rare to see it this intense." Her immediate reply was, "My Dad."

Mita, speaking for all three women, noted, "That notion of working hard was just part and parcel of who we were." The work ethic that Dr. Gupta's daughters inherited does not go unnoticed. One time, during a visit to their house, Vasudeva Uncle, as the girls call him, observed: "Just like your father, you guys all work very, very hard."

All three daughters made a point of acknowledging that the model of dedicated effort came not just from their father but also their mother. Mita explained,

> She worked very hard in establishing and running our home, getting up every morning and making our lunches to go to school—fresh sandwiches—and always having dinner made. And when we got home, there were homemade snacks waiting for us. And she didn't outsource much when we were growing up. She managed all of the infrastructure of a full household.

Most importantly, Alka, Mita, and Nisha all imbibed their father's values about education, agreeing that the number one lesson from him was, "Knowledge is power." All three women are ivy-league educated, not only because of their own determination and intelligence, but very much due to Dr. Gupta's emphasis on education and their parents' willingness to sacrifice to enable their daughters' chosen career paths. Again, this harkens back to the values Dr. Gupta took away from his childhood in Purkazi. Both Alka and Mita attended the prestigious Wharton School of Business at the University of Pennsylvania—Mita for undergraduate work, Alka for graduate work. Nisha attended Columbia University for her undergraduate degree and then earned her graduate degree in public health at Yale University. Nisha commented that for her college choices, she was instructed to "apply where you want, but pick a place with strong sciences and a strong graduate program," adding that Dr. Gupta believes that a strong graduate program impacts the quality of the undergraduate education.

Dr. Gupta recognizes his daughter's achievements as one of his most important contributions.

> I am most proud about being able to instill in my children the idea that they should pursue knowledge, which is what is most important, I think. And they did that. I very much think so. I was able to instill that, and they did as I did.

In His Own Words

On the plains of Northern India, during the first millennia, the heroic warrior Arjuna stood in the midst of the battlefield, trying to understand his duty. In this chaotic scene of elephants, warriors, and weapons, described in the Indian classic, *The Bhagavad Gita,* Lord Krishna advises Arjuna about the meaning of *dharma,* often translated as duty. In chapter 3, *The Yoga of Action,* Arjuna hears these words: "Perform your obligatory duty, because action is indeed better than inaction... Therefore, always perform your duty efficiently and without attachment to the results." Fast forward two thousand years. Mrs. Gupta describes how her husband Arjun, a namesake of that great warrior, worked hard, and always said: "This is my duty. This is it." She added, "He just kept doing what was right."

In the *Bhagavad Gita,* and the larger epic that contains it, *The Mahabharata,* Arjuna overcomes many inner obstacles to pursue his *dharma,* his duty, learning to persevere no matter how difficult things seemed. Arjun Gupta handles adversity similarly. His nephew Pankaj said,

I think Chachaji was well named. He's an embodiment of that kind of person. A warrior is someone who's just not going to give up, not only at one moment on the battlefield but basically in life. Dr. Gupta didn't give up. He just does it.

When asked what he learned from his life, Dr. Gupta responded,

Life is nothing but hard work. Each time we come across an adversity, don't give up. The inspiration that you get from your parents, or your brothers and sisters and others, is very important. I did have dilemmas sometimes, hard choices, but what's the alternative? There was no alternative in my mind. Arjuna had Krishna to depend on, but in my case, I think my parents were the guiding force. I could invoke them in my memory and thoughts of them helped.

Arjun Gupta's life has been built around dedication to duty. His life and work have impacted hundreds of people; he has guided and mentored two generations of students and relatives. Finally, after a career of almost fifty-five years, Dr. Gupta officially retired from Bowling Green State University in 2015, replete with awards, accolades, and global recognition. But it is difficult to suddenly stop doing what you have given your life to, and next to impossible to replace that passion. For Dr. Gupta, it was hard not to have a scheduled daily rhythm, going to work, and helping students. He shared, "I miss it." Yet he reflected,

It's been quite a life. A hard life, not an easy life. The most important lesson I learned was how to survive in different situations. It has not been easy, but I survived.

Of course, he did more than survive; he also thrived, which he acknowledges.

> As far as my career, I'm pretty satisfied. I had over twenty books written, which I wouldn't have done otherwise. And I have done okay.

No resume or biography can ever fully capture Arjun Gupta, since there is not just one Arjun Gupta. This one individual is made of many—the father to his children, the husband to his wife, the mentor to his students, the statistician to his colleagues, the guest to his international hosts, the grandfather, the uncle, the devoted brother, the lifelong friend, and more. Yet these multiple dimensions of his life and service weave a rich tapestry of this man who has touched so many people's lives, contributed enormously to the field of statistics, and inspires admiration in those who meet him. Dr. Gupta's own words best capture his view on life:

> I would tell others to be brave and accommodate according to circumstances. Be flexible. To the future generations, I would say, "Education, Education, Education."

In closing, we return to the *Bhagavad Gita*, where Arjuna is told that the highest of all mantras is the Gayatri mantra. As noted earlier, this same Gayatri mantra was assiduously transmitted to his daughters when Arjun Gupta drove them to school each day. Over the course of his life, he has lived by the principles that are embodied in this mantra—determination, noble actions, and an illumined intellect. The mantra translates as: "We meditate on the splendor of that Being who has produced this universe; may He enlighten our minds."

May Arjun Gupta be an inspiration for many generations to come.

Appendix A: Editorial Activities

(Listed in reverse chronological order of Dr. Gupta's participation from 1977 to 2014)

Editorial Board, *Journal of Advances in Mathematics*
Editorial Board, *Journal of Statistics in Medical Research*
Editorial Board: *Austin Statistics*
Editor, *International Journal of Business and Statistical Analysis*
Advisory Editor, *JSM Mathematics and Statistics*
Advisory Editor, *PanAmerican Mathematical Journal*
Advisory Editor, *E-STAT*
Editorial Board, *Journal of Statistical and Econometric Methods*
Associate Editor, *International Journal of Innovative Management, Information and Production*
Member Editorial Board, *Data Science*
Advisory Board, *Journal of Algebraic Statistics*
Advisory Board, *Journal of the School of Business Administration*
Editor, *Journal of Probability and Statistics*
Advisory Editor, *European Journal of Pure and Applied Mathematics*
Series Editor, *Statistics Textbooks and Monograph*, Chapman & Hall/CRC
Editorial Advisor, *Journal of Probability and Statistical Science*
Associate Editor, *Test*
Editor, *International Journal of Theoretical and Applied Mathematics*
Associate Editor, *Revista Mathematica*, UCM
Associate Editor, *Journal of Applied Statistical Science*
Associate Editor, *Journal of Intelligent Technologies and Applied Statistics*
Associate Editor, *Ohio Journal of Science*
Associate Editor, *Random Operators and Stochastic Equations*

Associate Editor, *Communications in Statistics*
Associate Editor, *Journal of Statistical Planning and Inference*
Editor (U.S.A.), *Statistical Theory and Methods Abstracts*
Member, International Editorial Board, *Communications in Statistics*
Member, Board of Editors, *ABACUS*
Editorial Consultant, *Mathematical Reviews*
Member, Board of Editors, *Creations in Mathematics*
Collaborator for *Mathematical Reviews* and *Zentralblatt für Mathematik*
Initiator for *Technical Report Series,* Department of Mathematics and Statistics, BGSU

Appendix B: Dr. Gupta's PhD Students

Sources:
https://www.bgsu.edu/content/dam/BGSU/college-of-arts-and-sciences/mathematics-and-statistics/documents/phd-graduates-list/phd-grads.pdf

https://genealogy.math.ndsu.nodak.edu/id.php?id=7004

Year	Name	Title	Accepted Position With:
1980-81	Dale Stanley Borowiak	Weighted Model Discrimination When the Errors are Multivariate Normal	Akron University
1981-82	Olawoye Soladoye Adegboye	On Testing Against Restricted Alternatives	University of Nigeria
	Jen Tang	Exact Distribution of Certain General Test Statistics in Multivariate Analysis	Bell Communication Research Lab
1982-83	Walfredo Ramos Javier	On the Distributions of Certain Random Matric Variates and Their Functions	University of North Dakota
1986-87	Terrence P. Logan	Discriminant Analysis Using Multiple Observations	University of Toledo
1988-89	Sam Ofori-Nyarko	Improved Estimation of the Covariance Matrix, the Precision Matrix and the Generalized Variance	University of Wisconsin

Year	Name	Title	Accepted Position With:
1989-90	Alphonse K. A. Amey	Robustness Study of Certain Multivariate Test Criteria When Sampling from a Contaminated Normal Model	Bowling Green State University; Botswana International University of Science & Technology
	Chi-Chin Chao	Inference About Covariance Matrices Under Repeated Measurements Model	Jacksonville State University
1990-91	Tamás Varga	Matrix Variate Elliptically Contoured Distributions: Stochastic Representation and Inference	Rose-Hulman Institute of Technology
1992-93	Bruce Johnson	Asymptotic Tests for the Equality of Several Correlation Matrices	National Family Opinion
1993-94	Danhong Song	Generalized Spherical and Liouville Distributions	Ohio Northern University
1995-96	Jie Chen	Inference About the Change Points in a Sequence of Gaussian Random Vectors Using Information Criterion	University of Missouri-Kansas
	Yining Wang	Characterizations Based on Conditional Structure and Their Statistical Application	University of Rochester
1996-97	Grzegorz Rempala	Limit Theorems for Random Permanents and U-Statistics of Infinite Order with Applications to Statistical Inference	University of Louisville

APPENDIX B - DR. GUPTA'S PHD STUDENTS

Year	Name	Title	Accepted Position With:
1997-98	Kaleli P. Asoka Ramanayake	Epidemic Change Point and Trend Analyses for Certain Statistical Models	University of Wisconsin-Oshkosh
1999-2000	John Carson	General Multivariate Hypothesis Testing Using Extensions of Hotelling's T-Squared	OHM Corporation
2003-04	Jose (Joel) Sanqui	Characterization and Statistical Inference for the Skew-Normal Distribution	Appalachian State University
2004-05	Solomon Harrar	Linear Models Under Non-Normality	South Dakota State University
	Jin Xu	Robustness Study of Some Multivariate Tests in Generalized Linear Models	Univ. of California-Riverside
2005-06	Keshav Jagannathan	Statistical Inference and Goodness-of-fit tests for Skewed Double Exponential Models	Coastal Carolina University
2006-07	Dhanuja Kasturiratna	Assessing the Distributional Assumptions in One-Way Regression Model	Northern Kentucky University
2008-09	Deniz Akdemir	On a Class of Multivariate Skew Distributions: Properties and Inferential Issues	StatGen Consulting
2010-11	Ngoc Nguyen	Estimation of Technical Efficiency in Stochastic Frontier	University of Kentucky
2011-12	Mohammad Aziz	Study of Unified Multivariate Skew Normal Distribution with Applications in Finance and Actuarial Science	University of Bangladesh

Year	Name	Title	Accepted Position With:
2013/14	Abeer Hasan	A Study of Skew T Distribution with Application	Humboldt State University
2015/16	Ying-Ju Chen	Jackknife Empirical Likelihood and Change Point Problems	University of Dayton, Ohio
2017/18	Doaa Basalamah	Statistical Inference for a New Class of Skew T Distribution and its Related Properties	Umm Al-Qura University, Mecca, Saudi Arabia
	Ramadha Dilhani Piyadi Gamage	Empirical Likelihood for Change Point Detection and Estimation in Time Series Models	Western Washington University, Bellingham, WA
2018/19	Amani Alghamadi	Study of Generalized Lomax Distribution and Change Point Problem	King Abdulaziz University, Jeddah, Saudi Arabia

APPENDIX C: RESEARCH COAUTHORS

Sources:
https://www.researchgate.net/profile/Arjun_Gupta2
https://link.springer.com
https://scholarworks.bgsu.edu/math_stat_pub/

Deniz Akdemir	StatGen Consulting
Mohammad A. Aziz	Pakistan
Jim Albert	Bowling Green State University
Amani Alghami	Saudi Arabia
Alphonse Kwame Adjei Amey	Botswana International University of Science and Technology
Doaa Basalamah	Saudi Arabia
Asit Basu	University of Missouri, Columbia
Olha Bodnar	Stockholm University
Taras Bodnar	Stockholm University
Hamparsum Bozdogan	University of Tennessee
Jie Chen	Augusta University
John T. Chen	Bowling Green State University
Ying-Ju Chen	University of Dayton
J. Armando Dominguez-Molina	Universidad Autónoma de Sinaloa
Yasunori Fujikoshi	Hiroshima University

Ramadha D. Piyadi Gamage	University of Colombo, Sri Lanka
Graciela González Farías	Centro de Investigación en Matemáticas (CIMAT)
Yasunori Fujikoshi	Hiroshima University
Vyacheslav L. Girko	University, Kyiv, Russia
Solomon W. Harrar	University of Kentucky
Dominique Haughton	Bentley University
W. R. Javier	University of the Philippines
Anwar H. Joarder	Northern University of Business & Technology Khulna; University of Liberal Arts Bangladesh (ULAB)
D. G. Kabe	University of Halifax
Dhanuja Kasturiratna	Northern Kentucky University
Genshiro Kitagawa	Research Organization of Information and Systems (ROIS)
Sadanori Konishi	Kyushu University
S. Kotz	University of Maryland
Ankush Kumar	India
Debasis Kundu	Indian Institute of Technology Kanpur
Baokun Li	Memorial University of Newfoundland, Canada
T. P. Logan	Private Consultant
T. N. Mehrotra	India
Raúl Alejandro Morán-Vásquez	University of Antioquia

Saralees Nadarajah	The University of Manchester
Daya Nagar	University of Antioquia
Naoto Niki	Institute of Statistical Math, Japan
Diem M. Nguyen	Hanoi University
Truc Nguyen	Bowling Green State University
Wei Ning	Bowling Green State University
Solomon Sarkodie Ofori	Water Resources Commission
Tohru Ozaki	Japan
Leandro Pardo	Complutense University of Madrid
Asoka Ramanayake	University of Colombo
V.K. Rohatgi	Bowling Green State University
Alejandra Roldan-Correa	Mexico
Luz Estela Sánchez	University of Antioquia
Jose Almer Sanqui	Appalachian State University
Amadou Sarr	Sultan Qaboos University
Stanley Sclove	University of Illinois at Chicago
Yo Sheena	Shinshu University
D. Song	Peking University
Saowanit Sukparungsee	Thailand
Bon Sy	CUNY Graduate Center
Gabor J. Szekely	National Science Foundation Alexandria VA
Kunio Tanabe	Waseda University
Mahdi Teimouri	Gonbad Kavous University

Tomas Varga	Actuary, Hungary
Tonghui Wang	New Mexico State University
Yining Wang	Bowling Green State University
Jacek Wesolowski	Warsaw University of Technology
Yanhong Wu	California State University Stanislaus
Jin Xu	Purdue University
Rendao Ye	Purdue University
Wei-Bin Zeng	University of Pittsburgh

APPENDIX D: PUBLISHED BOOKS

Bozdogan, H., and Arjun K. Gupta, eds. *Multivariate Statistical Modeling and Data Analysis.* Dordrecht, Netherlands: D. Reidel Publishing Co., 1987.

Bozdogan, H., and Arjun K. Gupta, et al., eds. *Theory and Methodology of Time Series Analysis, Vol. 1.* Dordrecht, Netherlands: Kluwer Academic Publishers, 1994.

Bozdogan, H., and Arjun K. Gupta, et al., eds. *Multivariate Statistical Modeling, Vol. 2.* Dordrecht, Netherlands: Kluwer Academic Publishers, 1994.

Bozdogan, H., and Arjun K. Gupta, et al., eds. *Engineering and Scientific Applications of Informational Modeling, Vol. 3.* Dordrecht, Netherlands: Kluwer Academic Publishers, 1994.

Chen, Jie, and Arjun K. Gupta. *Parametric Statistical Change Point Analysis: With Applications to Genetics, Medicine, and Finance. 2nd ed.* New York, NY: Birkhäuser, 2012

Gupta, Arjun K., ed. *The Analysis of Categorical Data.* Special Issue of *Communications in Statistics—Theory and Methods*, Volume 12, Number 11. New York: Marcel Dekker, 1983.

Gupta, Arjun K., ed. *Advances in Multivariate Statistical Analysis.* Dordrecht, Netherlands: D. Reidel Publishing Co., 1987.

Gupta, Arjun K. and V. L. Girko, eds. *Multidimensional Statistical Analysis and Theory of Random Matrices.* Utrecht, Netherlands: VSP, 1996.

Gupta, Arjun K. and D. G. Kabe. *Theory of Sample Surveys.* Singapore, Malaysia: World Scientific Publishing, 2011.

Gupta, Arjun K. and D. G. Kabe. *Design and Analysis of Experiments*, Singapore, Malaysia: World Scientific Publishing, 2013.

Gupta, Arjun K. and S. Nadarajah, eds. *Handbook of Beta Distribution and Its Applications*. New York: Marcel Dekker, 2004.

Gupta, Arjun K. and D. K. Nagar. *Matrix Variate Distributions*. Boca Raton, FL: Chapman & Hall/CRC, 2000.

Gupta, Arjun K. and Tomas Varga. *Elliptically Contoured Models in Statistics*. Dordrecht, Netherlands: Kluwer Academic Publishing Group, 1993.

Gupta, Arjun K. and Tomas Varga. *An Introduction to Actuarial Mathematics.*, Dordrecht, Netherlands: Kluwer Academic Publishers, 2002.

Gupta, Arjun K., Tomas Varga, and T. Bodnar. *Elliptically Contoured Models in Statistics and Portfolio Theory*, New York: Springer, 2013.

Gupta, Arjun K., Wei-bin Zeng, and Yanghon Wu. *Probability and Statistical Models: Foundations for Problems in Reliability and Financial Mathematics*. Boston, Ma: Birkhäuser, 2010.

Kabe, D. G. and Arjun K. Gupta. *Experimental Designs: Exercises and Solutions*. New York: Springer, 2007.

Martínez-ramón, Manuel, Arjun Gupta, et al. *Machine Learning Applications in Electromagnetics and Antenna Array Processing*. Artech House: Boston, Mass: 2020

Nagar, D. K. and Arjun K. Gupta. *Contributions to the Complex Matrix Variate Distribution Theory*. Medellín, Colombia: Universidad de Antioquia, 2009.

Sy, Bon K. and Arjun K. Gupta. *Information-Statistical Data Mining*. Boston, MA: Kluwer Academic Publishers, 2004.

Appendix E: International Assignments

1974: Ontario, Canada (Windsor)
1979: Iraq (Basra)
1979 Rumania (Bucharest, Transylvania)
1980: Brazil (Campinas)
1983: India (Rajasthan)
1984 Argentina
1985: Spain (Bilbao)
1987: Ghana (United Nations Statistical Consultant) Fall
1987: Togo (Lome)
1988 Poland (Warsaw)
1990: India (Delhi)
1997: China
1999: Colombia (Medellín)
1999: South Africa (Durban, Cape Town, Stellenbosch, Bloemfontein, Pretoria)
2000: Taiwan (Taipei, Kaohsiung)
2000: Mexico (Guanajuato)
2001: Mexico (Puerto Marelos)
2002: Kuwait
2002: Japan (Tokyo, Hiroshima, Kyoto)
2002: China (Beijing, Shanghai)
2003: Switzerland (Neuchâtel)
2004: Taiwan
2005: Turkey (Kayseri)
2007: Malaysia (Kuala Lumpur)
2007: Portugal (Lisbon)
2008: Turkey (Kayseri, Cappadocia)
2008: Thailand (Bangkok)
2009: Taiwan
2009: Vietnam (Ho Chi Minh City and Hanoi)

2010:   Egypt (Cairo)
2011:   Thailand (Bangkok)
2011:   Mexico (Puerto Marelos)
2012:   Saudi Arabia (Dhahran)
2013:   Cameroon (Buea)
2013:   Thailand
2013:   Sudan
2013:   Tobago/Trinidad
2014:   Germany (Berlin)
2015:   Thailand
2016:   Turkey

Appendix F: End Notes

Prologue

Page
1 *Dr. Pillai.*: https://www.stat.purdue.edu/giving/pillai.html

1 *Chachaji:* A respectful term for "uncle" and specifically, one's father's younger brother.

2. The Companionship of Siblings

Page
19 *Dr. Pradeep Gupta*: https://everythingsouthcity.com/2016/12/new-mayor-and-vice-mayor-sworn-in-at-south-san-francisco-city-council-reorganization-event/

3. Knowledge is Power

Page
36 *S.N. Singh*: https://www.bhu.ac.in/science/statistics/

36 *Sir R. A. Fisher*: https://en.m.wikipedia.org/wiki/Ronald_Fisher

36 *Professor C. R. Raos*: http://mathematics.ceu.edu/rao

37 *V. S. (Vasant Shankar) Huzurbazar*: https://en.wikipedia.org/wiki/V._S._Huzurbazar

41 *Sharadchandra Shankar Shrikhande*: https://en.m.wikipedia.org/wiki/Sharadchandra_Shankar_Shrikhande

4. In Pursuit of a PhD

Page
63 *Aridaman Jain*: https://www.waldenu.edu/why-walden/faculty/aridaman-jain

68 *In 1963, the Department of Statistics:* For a detailed history of statistics at Purdue, see the essay *Purdue Statistics: A Journey Through Time* in the book *Strength in Numbers: The Rising of Academic Statistics Departments in the U. S.* published by Springer. https://www.springer.com/us/book/9781461436485

Page
68 *Shreeram Shankar Abhyankar:* https://en.wikipedia.org/wiki/Shreeram_Shankar_Abhyankar

71 *a paper with him on the multivariate test, which was published in a journal:* "On the distribution of the second elementary symmetric function of the roots of a matrix." 1967

5. Post-Graduation: New Family, New Locations

Page
86 *"On the Exact Distribution of Wilks' Criterion" and "Noncentral distribution of Wilks' statistic in MANOVA":* MANOVA is an acronym for multivariate analysis of variance. https://projecteuclid.org/journals/annals-of-mathematical-statistics/volume-42/issue-4/Noncentral-Distribution-of-Wilks-Statistic-in-Manova/10.1214/aoms/1177693238.full

90 *University of Michigan:* https://lsa.umich.edu/stats/about/history.html

91 *He became a consultant for Mathematical Reviews:* http://www.ams.org/publications/math-reviews/math-reviews

95 *Vijay Rohatgi:* Invited paper for the International Conference on Optimization in Statistics, December 1971, Bombay, India

95 *Dr. Zakkula Govindarajalu*: https://www.legacy.com/obituaries/Kentucky/obituary.aspx?page=lifestory&pid=147057831

95 *Mrs. Gayatri Govindarajalu*: https://www.legacy.com/obituaries/kentucky/obituary.aspx?n=gayatri-govindarajulu&pid=163700203

96 *I was a parametric statistician:* Parametric statistics is a branch of statistics which assumes that sample data comes from a population that can be adequately modeled by a probability distribution with a fixed set of parameters. Conversely a non-parametric model makes no assumptions about a parametric distribution when modeling the data. (Source: Wikipedia).

96 *Dr. A. P. Basu*: https://imstat.org/2016/02/17/obituary-asit-basu-1937-2015/

98 *Subhash C. Goel*: http://faculty-history.dc.umich.edu/faculty/subhash-c-goel/memoir?quicktabs_1=2

99 *Eugene Lukacs*: https://www.bgsu.edu/arts-and-sciences/mathematics-and-statistics/faculty-and-staff/retired-and-former-faculty/eugene-lukacs.html

101 *the initiative to establish a PhD program for statistics:* https://www.bgsu.edu/arts-and-sciences/mathematics-and-statistics/department-history/history-1959-1973.html

103 *Dr. Albert:* https://www.bgsu.edu/news/2018/04/distinguished-university-professor.html; Source of all quotations by Dr. Albert are transcriptions of remarks from Dr. Gupta's retirement party

104 *John Chen.* Source of all quotations by Dr. Chen are transcriptions of remarks from Dr. Gupta's retirement party

104 *Lukacs Distinguished Visiting Professor*: https://en.wikipedia.org/wiki/Lukacs_Distinguished_Professor

6. Professor Gupta

Page

105 *Brad Efron:* https://en.wikipedia.org/wiki/Bradley_Efron

105 *Persi Diaconis*: https://en.wikipedia.org/wiki/Persi_Diaconis

108 *Dr. Jen Tang*: https://www.krannert.purdue.edu/directory/bio.php?username=jtang

115 *Professor Cas Troskie*: https://www.news.uct.ac.za/article/-2010-03-29-passing-of-an-era-tribute-to-cas-troskie

7. A Scholar with Heart

Page

140 *Harvard looked down on creating a Department of Statistics*: https://statistics.fas.harvard.edu/history

145 *Schatzoff: Exact Distribution of Wilks' Likelihood Ratio Criterion and Comparisons with Competitive Tests,* Martin Schatzoff, 1964

146 *the first ones to derive the distribution of the test statistic in closed form:* https://www.researchgate.net/publication/ 243082206_On_the_Exact_Distribution_of_Wilks's_Criterion

149 *the book with Dr. Jie Chen on change point analysis:* Chen, Jie, and Arjun K. Gupta. *Parametric Statistical Change Point Analysis: With Applications to Genetics, Medicine, and Finance.* 2nd ed. New York, NY: Birkhäuser, 2012

150 *Dr. Gupta's 529 publications listed on Researchgate.net show a total of 46,280 reads and 7,224 citations:* https://www.researchgate.net/ profile/Arjun-Gupta-9

152 *A headline in the New York Times in 2012:* https:// bits.blogs.nytimes.com/2012/01/26/what-are-the-odds-that-stats-would-get-this-popular

152 *An article on the Harvard Department of Statistics website*: https:// statistics.fas.harvard.edu/history

8. International Ambassador of Statistics

Page
186 Dr. Hamparsum Bozdogan: https://haslam.utk.edu/experts/ hamparsum-bozdogan

www.ingramcontent.com/pod-product-compliance
Lightning Source LLC
Chambersburg PA
CBHW061149170426
43209CB00035B/1953/J